SpringerBriefs in Applied Sciences and Technology

Manufacturing and Surface Engineering

Series editor

Joao Paulo Davim, Aveiro, Portugal

More information about this series at http://www.springer.com/series/10623

Manjuri Hazarika · Uday Shanker Dixit

Setup Planning
for Machining

 Springer

Manjuri Hazarika
Department of Mechanical Engineering
Assam Engineering College
Guwahati
India

Uday Shanker Dixit
Department of Mechanical Engineering
Indian Institute of Technology
Guwahati
India

ISSN 2191-530X   ISSN 2191-5318   (electronic)
SpringerBriefs in Applied Sciences and Technology
ISBN 978-3-319-13319-5   ISBN 978-3-319-13320-1   (eBook)
DOI 10.1007/978-3-319-13320-1

Library of Congress Control Number: 2014955200

Springer Cham Heidelberg New York Dordrecht London

Printed on acid-free paper

Springer International Publishing AG Switzerland is part of Springer Science+Business Media (www.springer.com)

Preface

Manufacturing is the oldest process responsible for producing various types of products involving all spheres of human civilization. With the growing demand for variety in products, customization and quality, the need arises to explore innovative ways to meet these challenges. Process planning/setup planning plays a key role in manufacturing process and it essentially paves the way from design phase to manufacturing phase. It bridges the functional gap between design and manufacturing. Process planning and setup planning methods are vital for the functionality, efficiency, cost, and quality of a finished product. It has been an active research area in the last three decades and widely investigated by various researchers.

The scope of this book is limited to setup planning in machining context. Setup planning is an intermediate phase of process planning and it is essentially the core of a process planning system. Setup planning includes determination of the setups needed to machine a component, setup and machining operation sequencing, selecting datum, and selecting jigs and fixtures. The knowledge of different types of features, their dimensions, tolerances, machine tools and their capabilities, cutting tools, machining operations, and fixtures are essential for setup planning. Challenging issues like automation, integration, compatibility and proper interfacing, flexibility in setup planning, etc., are addressed in this book. The use of soft computing techniques in solving setup planning problems is also discussed.

The first chapter of the book essentially introduces process planning and setup planning in machining context along with the different approaches of setup planning. In Chap. 2, different phases of setup planning task, viz, feature grouping, setup formation, datum selection, machining operation sequencing, and setup sequencing are discussed with relevant examples. Setup planning has been an active area of research for a long time. Chapter 3 reviews major efforts of setup planning by various researchers using diverse methods. Application of the traditional approaches like decision tree, decision table, group technology, algorithms and graphs, artificial intelligence tool like expert system, soft computing techniques like fuzzy sets, neural networks, and evolutionary optimization methods to setup planning are presented in this chapter. Chapter 4 describes the application of fuzzy set theory to take care of the uncertainty and imprecision associated with setup planning knowledge. Chapter 5 addresses an important issue of assigning proper membership grades to fuzzy variables. A method for fine-tuning the membership grades combining the expert's opinion and available practical data is described with an example. Chapter 6 reviews different types of fixtures and the relevant research in the area. Emphasis is given on the need for fixturing consideration during setup planning stage for a practical and feasible setup planning solution.

This book may be used as a part of a course on manufacturing engineering at both the undergraduate and postgraduate level. It can also be used as a reference by the researchers in the broad area of process planning and setup planning. We welcome the feedback of readers.

We thank Prof. J. Paulo Davim for motivating to write us a monograph on setup planning. We also want to acknowledge Dr. Sankha Deb for the fruitful discussions we had with him on setup planning. The cooperation of the staff members of Springer is also acknowledged.

Guwahati, India   Manjuri Hazarika
                                                                                   Uday Shanker Dixit

Contents

Chapter 1
Process Planning in Machining

Abstract In this chapter a brief introduction to process planning and setup planning in machining is described. Computer aided process planning/setup planning, its various approaches and main constraints are discussed. Some important concepts in relation to process planning and setup planning, viz. features and their tool approach directions, idea of datum, part dimensions and tolerances are presented. A brief discussion about flexibility in setup planning and data exchange formats is also included. Differences in setup planning for prismatic as well as rotational parts and setup planning for green machining are presented briefly.

Keywords Process planning · Setup planning · Tool approach direction (TAD) · Datum · Prismatic and rotational part

1.1 Introduction

Process planning is an important activity in discrete part manufacturing. Process planning maps the design information of a part to systematic manufacturing steps through which raw stock is converted into finished product. It is the post-design and pre-manufacturing activity that bridges the functional gap between design and manufacturing. Process planning can be defined as the act of preparing detailed manufacturing instructions to produce a part with the available resources at the lowest possible cost and of the best quality.

In today's competitive market, integration among different functional areas of a manufacturing industry is essential for improvements in quality, efficiency, cost and time. Each of design, production planning, manufacturing, quality control and other support functions act as a part of a unified system rather than a stand alone system. Therefore, the need arises to integrate design and manufacturing phases to allow a path from initial concept to a finished product. This need is fulfilled by process planning that brings together design and manufacturing as shown in Fig. 1.1.

© The Author(s) 2015
M. Hazarika and U.S. Dixit, *Setup Planning for Machining*,
SpringerBriefs in Manufacturing and Surface Engineering,
DOI 10.1007/978-3-319-13320-1_1

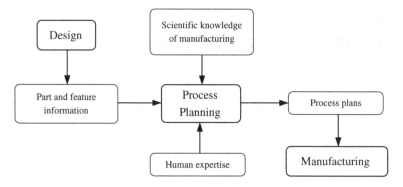

Fig. 1.1 Process planning: the bridge between design and manufacturing

There are several key issues and challenges to be addressed for successful development and implementation of efficient process plans. Automation is gaining prime importance in modern manufacturing industries to fulfill the need for improved productivity and quality. Automating process planning is the key for seamless integration between computer aided design (CAD) and computer aided manufacturing (CAM). Many sophisticated CAD and CAM systems are available commercially. However, equivalent commercial automatic process planning systems are sparse. Another important issue in integration of automated process planning with CAD/CAM is the incompatibility of the equipment and software. Development of proper interface standards is necessary for exchange of information among design, process planning and manufacturing stages. Manufacturing industry is faced with the challenges of product variety and customization combined with the requirement of enhanced product quality at lower cost. Attaining the specified design tolerances is a key factor for the quality as well as the functionality of a component. Moreover, ensuring that the product meets design specifications and is of required quality is not enough. The product should be cost-effective, and it should be completed in time. It is the task of process planning to ensure that these requirements are met. The manufacturing industries have to innovate ways to reduce the time taken to design, manufacture and market the product. Through proper process planning, it is ensured that the product is made to the correct specifications, at the lowest possible cost and completed on time. Another important consideration in process planning is flexibility. Adaptation to the changing scenario is a crucial factor in this era of lean and agile manufacturing. There is continuous improvements and redesigning of products to meet the customer's demand. Therefore the process plans have to be flexible and ready to adapt to the changes in design as well as manufacturing environment. Thus, process planning plays a vital role in determining the manufacturing cost, quality of the product, manufacturing lead time and efficiency of the production system.

There are two types of process planning in manufacturing—machining process planning and assembly process planning. In this book, machining process planning, in particular setup planning for machining a component is covered. The task of machining a component is broken down into a number of individual process steps. The constituents of machining process planning can be decomposed into several sequential phases. Each phase is dependent on its previous phase and the different tasks to be performed within each phase are interrelated. As per Xu et al. [43], the phases are as follows:

Design interpretation: interpretation of the part design and analysis of the requirements of the finished part regarding features to be machined, design dimensioning, various tolerances, surface finish, selection of the raw stock and materials to be used;

Process selection and machine selection: selection of machining processes for producing the features of the part, analysis of the process capabilities of the processes, selection of machine tools and cutting tools;

Setup planning: determination of the setups needed to machine the part, setup and operation sequencing, selecting/designing jigs and fixtures;

Process parameter selection: calculation of speed, feed and depth of cut for each machining operation;

Cycle time estimation and scheduling: estimation and balancing of machining cycle time in each setup and scheduling of jobs for shop floor production;

Cost evaluation: economic analysis of manufacturing cost and quality evaluation;

Documentation: documentation of process plan.

A process plan has to be detailed in respect of machining processes, process sequence, machine tools, cutting tools, fixtures, locating and clamping, setup datum, process parameters, etc. A process planner should be capable of interpreting design information, possess knowledge of manufacturing processes and their relative costs and process capabilities. Moreover he has to be familiar with the resources in the shop.

1.2 Setup Planning: Core of Process Planning

Setup planning constitutes the core of a process planning system. It is an intermediate phase of process planning. Setup planning is the act of preparing instructions for setting up parts for machining. A setup is basically the way in which the part is oriented and fixtured in one particular position in the machine tool for machining. As soon as the part's position is changed, it is considered a new setup. A group of features are machined in a setup without repositioning the part.

Setup planning consists of the following steps:

- Feature grouping
- Setup formation
- Datum selection
- Machining operation sequencing and setup sequencing
- Selecting/designing jigs and fixtures

Fig. 1.2 A setup planning
system

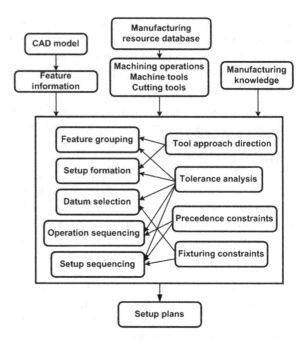

These steps are discussed elaborately in Chap. 2. Figure 1.2 describes the frame-work for an ideal integrated setup planning system. It takes information on fea-tures of a part, machining operations, machine tools and cutting tools as inputs. Based on these inputs, manufacturing knowledge, and constraints in setup plan-ning such as tool approach direction, precedence constraint and so on (discussed in Sect. 1.4), setup planning is done. Feature grouping, setup formation, datum selection, machining operation and setup sequencing tasks are performed and complete setup plans are formulated.

Interpretation of the part design is one of the major factors of setup plan-ning. The input to setup planning is the part specification and its output is the manufacturing instructions. The part representation database comprises the input containing information of the part including features of the part, part dimensions, shape, tolerances, surface finish, etc. The planner has to extract this information in order to select the necessary machining processes to machine the part. Some fundamental topics that are crucial for setup planning are discussed hereunder.

1.2.1 Features of a Part

Features are the medium for transfer of information in the CAD/CAM integra-tion. A part may contain different features. A feature is a specific geometric shape formed on the surface, edge or corner of a part. A machining feature is generated by a machining process. The machining features represent the geometry of a part. There are many definitions for machining feature. Gao and Shah [14] defined it as a continuous volume that can be removed by a single machining operation in a single setup. Tseng and Joshi [37] defined it as a portion of a part having some manufacturing significance that can be created with certain machining operations. According to Yan et al. [46] machining feature is a distinctive object in a workpiece with geometric and topological characteristics. Features are considered as main factors in design and manufacturing integration as various design and manufacturing data can be associated with a feature. Features play an important part in creating solid models. Some examples of machining features are face, hole, slot, step, pocket, chamfer, etc. Different machining operations such as turning, milling, drilling, forming are used to generate features. Figure 1.3 shows different types of features found in a part.

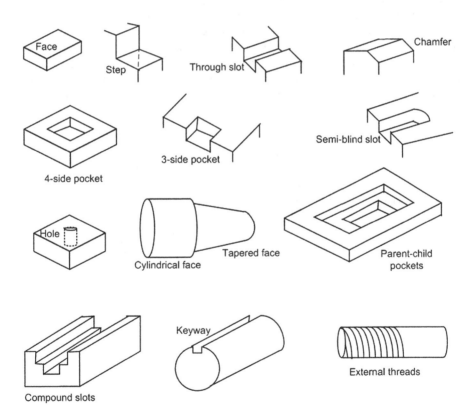

Fig. 1.3   Different types of features

1.2.2 Part Dimensions

Dimensions are the numerical values associated with the size of a part. Geometric Dimensioning and Tolerancing (GD&T) is a universal language that allows a designer to precisely represent part dimensions in a drawing. Dimensions are given in design drawings using lines, symbols and notes. The process planner has to extract the different dimensions of the part features from the part drawing. In manual part drawing, the draftsman makes a detailed drawing of the part showing all the three views, i.e. elevation, plan and end/side view. Using CAD systems, the designer can construct all these three views more accurately and easily. Once the part drawing is complete with all the three views, dimensions and tolerances of the part features are added in these views to fully describe a part. There are various methods of dimensioning a part. In Cartesian method, dimensions are given both in X and Y axis directions. In Ordinate method, same reference is used to give dimension in one direction. Moreover angular dimensions are used to show arcs and angles. Different dimensioning methods are shown in Fig. 1.4. Features should be dimensioned in that view where the true size and shape of the feature is shown. Basic dimensions establish the true position of a feature from datum features and between interrelated features. Unnecessary dimensioning should be avoided. In a properly dimensioned part drawing, there is no need to calculate the size and position of any feature at the manufacturing stage. CAD systems allow much flexibility in dimensioning methods regarding dimension type (Cartesian/ Ordinate), change in sizes, units, etc. Moreover, there is associability between the dimensions of the model and drawing in CAD systems. This is called parametric

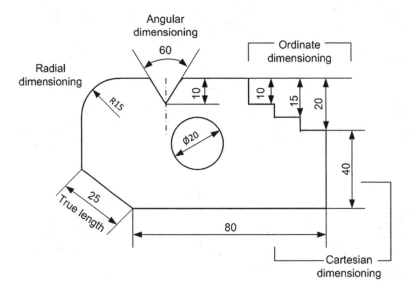

Fig. 1.4 Different types of dimensions

design which allows automatic update of any change made in model or drawing mode. For example, if a dimension of the part is changed in any of the three views in the drawing, it is automatically reflected in the model and vice versa. For process planning, dimensions and size and shape of a feature play a crucial role in selection of the manufacturing process, machine tools and cutting tools.

1.2.3 Tolerances

Manufacturing a part to its exact dimensions is an impossible task due to many limitations faced in practical situations. The dimensions may vary from the exact values actually needed. Tolerance determines the limits of this production variability. During design phase, tolerances are assigned to part dimensions to account for the variability in dimensions during production phase. Tolerance determines a range of acceptable values for the dimensions of the toleranced feature. Dimensions represent the required size of a feature and tolerances represent the required precision. Both engineering design and manufacturing persons are concerned with the tolerances specified on engineering drawings. Tolerances of the part features greatly influence the selection of machine tools and machining processes by process planners. Attaining the specified design tolerances is a key factor for the quality as well as the functionality of a component. It also greatly influences the manufacturing cost. If the tolerance is very small or tight, i.e. very less deviation is allowed in dimension, manufacturing cost goes up. Alternatively, if the tolerance is very large, high deviation is allowed and rejection rate is more thus adding to manufacturing cost. It is the task of process planning to select appropriate processes and machines to ensure that specified design tolerance requirements are met. Thus tolerance specification is an important criterion to be considered during process planning. It is very important for assembly operation and inspection phase as well.

In 1973, the American National Standards Institute (ANSI) introduced a system called Geometric Dimensioning and Tolerancing (GD&T) that establishes a uniform practice for incorporating dimensions and tolerances in engineering drawing. This system was referred to as ANSI Y14.5-1973. In 1994, after enhancement and modifications the standard has become ASME Y14.5-1994 which is a universal Standard for dimensioning and tolerancing [6]. Designers can precisely represent the positions, dimensions and tolerances of part features with the help of Geometric Dimensioning and Tolerancing (GD&T). Tolerances can be mainly classified into two types, dimensional tolerances and geometric tolerances. Dimensional tolerance is added in a feature with the maximum and minimum limits within which the dimension is acceptable. There are two ways of representing dimensional tolerance, unilateral tolerance and bilateral tolerance. In unilateral tolerance specification, tolerance is expressed either more or less from the basic size, e.g. $50^{+0.02}_{-0.00}$, $50^{+0.00}_{-0.04}$ or $50^{+0.04}_{+0.02}$ mm. The convention is to show the upper limit of the dimension always on the top. In bilateral tolerance specification, tolerance

Fig. 1.5 Example of
dimensional tolerance

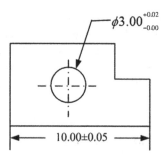

is expressed both ways (more and less) from the basic size, e.g. 50 ± 0.02 or $50^{+0.05}_{-0.01}$ mm. Although dimensional tolerance controls size to a certain limit, it is unable to control the form or shape of a feature. Figure 1.5 shows an example of adding dimensional tolerance in engineering drawing. The length of the part is acceptable in the range 9.95–10.05 mm. The hole-diameter should be in the range 3–3.02 mm.

Geometric tolerances specify the maximum allowed deviations in form or position of a feature from the true geometry. In effect, geometric tolerance controls the form of a feature by defining a tolerance zone within which the feature must be contained. Geometric tolerances can be further classified into 14 types as per ANSI Y14.5-1994. Table 1.1 shows different geometric tolerances with their ANSI symbols.

Proper tolerance specification in design phase is crucial for process planning as selection of machines, processes and cutting tools depends on the dimensions and tolerances of the part features. When all the tolerances are assigned to the different parts of an assembly, a tolerance analysis is to be performed to check for functionality and proper assembly of the mating parts. Dimensions and tolerances of part features can be incorporated in CAD files and those are to be extracted by the process planner along with other feature information. Thus tolerance information is an important issue both in design and manufacturing. Designers use tolerances among related design dimensions and manufacturers need tolerance information for process planning, inspection and quality control.

1.2.4 Datum

For creating reference for a component to be machined, datum is used. Datum may be some imaginary planes as shown in Fig. 1.6 or they may be some datum features comprising the component. Generally datum features rest on datum planes. Datum planes are not a part of the component geometry but it is a part of the component's reference geometry. Reference geometry is expressed in terms of datum planes/features. The imaginary plane on which a component lies during

Table 1.1 Geometric tolerances with their ANSI symbols

Tolerance category	Type of tolerance	ANSI symbols
Form tolerance	Straightness	
	Flatness	
	Circularity	
	Cylindricity	
Orientation tolerance	Parallelism	
	Perpendicularity	
	Angularity	
Location tolerance	Position	
	Symmetry	
	Concentricity	
Profile	Profile of a line	
	Profile of a surface	
Runout	Circular runout	
	Total runout	

machining is called the primary datum plane as shown in Fig. 1.6. The actual feature of the component (a face in this case) that lies on the primary datum plane is called the primary datum. Secondary datum plan is perpendicular to the primary datum plane and the tertiary datum plane is perpendicular to both the primary datum plane and secondary datum plane. Secondary and tertiary datum features lay on their respective datum planes. A datum feature may be a face, an axis, a curve or a point. Datum features are used for locating a component. In Fig. 1.6, six locators are used in three datum faces to locate the prismatic component. In case of rotational components, both holes and surfaces can be used as datum features.

In setup planning once the setups are formed, the setup datums are to be selected. The decision on selecting suitable datum for each setup depends on various factors like feature tolerance relationships, surface area of a face, its orientation, symmetry, and surface quality. Tolerance relations with other features and maximum area face are the most widely used criteria for selecting datum. To select datum for a setup in case of a prismatic part, first all the faces of the part are

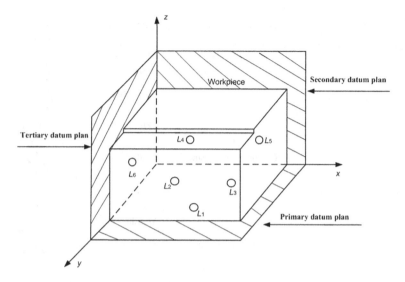

Fig. 1.6 Three datum planes for a prismatic part

identified. The faces having an orientation different from the faces being machined in that setup are sorted out. In case of rotational parts, generally the two faces perpendicular to the axis of the part are selected as datum.

1.3 Computer-Aided Setup Planning

Process planning/setup planning requires the creativity and analytical skills essential for design in addition to in-depth manufacturing knowledge needed for actual production. Process planning in discrete part manufacturing involves a number of decisions such as design interpretation, selection of material, selection of machining operations, machine tools and cutting tools, setup planning, selection of jigs and fixtures, determination of cutting conditions and so on. Setup planning is traditionally performed by experienced process planners based on his intuition and rules of thumb gained from his experience. However, manual planning has its own shortcomings. Manual planning is tedious, time consuming and prone to human errors. It is closely tied to the personal knowledge and experience of the process planner. This results in inconsistent plans due to variability in human perception. Since it is a highly skilled job done by expert process planners, there is a risk of losing the expertise as the process planner retires. Moreover, the modern manufacturing industry is undergoing profound changes with emphasis on automation for improved productivity and quality. With increase in demand for novelty and variety in products, standard computer aided design (CAD) and computer aided manufacturing (CAM) tools are used by the industries. Addressing these issues leads to the need of computer aided process planning/setup planning systems. In view of it, researchers have been trying to automate setup planning. Although the efforts to automate setup planning have been going on since 1980s, it is still a complex task. This is because an optimum setup plan is dependent on a number of factors such as type and shape of raw material, availability of machines and tools, quantity of production, cost, desired tolerances and policy of management. Setup plan for mass production with high production volume is different from setup plan for job shop production with low production volume. An experienced process planner evaluates these factors mentally and takes an appropriate decision based on intuition and experience. It is very difficult to capture and store the knowledge of an experienced process planner in the form of a computer code. Nevertheless, the ultimate goal of automating setup planning is to achieve the desired quality of the finished product at the lowest possible production cost and with the minimum manufacturing time. This ultimate goal affects all decision making in setup planning. Realisation of this ultimate goal may vary with different manufacturing settings. The benefits that can be derived from computer-aided process planning/setup planning are as follows:

- It bridges the functional gap between CAD and CAM.
- It enables concurrent engineering and builds a proper environment for computer-aided manufacturing.
- Overall manufacturing cost is reduced by reducing planning time, direct labour, etc.
- It generates detailed, complete and consistent plans.
- It reduces manufacturing lead time and enhances productivity.

1.4   Constraints in Setup Planning

A feasible and optimal setup plan involves a number of constraints and objectives that are mutually conflicting considering technological, geometrical and economical aspects of both design and manufacturing domain. Some constraints are very basic and of higher priority, which, if violated, will result in infeasible setup plans. Interactions among the features lead to precedence constraints which are also called hard constraints. For example, machining of datum and reference features first is a hard constraint. There are also soft constraints, which being violated, will result in inferior setup plans. For example, attaining a particular tolerance is a soft constraint. Different constraints applicable to various stages of setup planning are discussed in the following sections.

1.4.1   Tool Approach Direction (TAD) of a Feature

Tool approach direction (TAD) of a feature is the unobstructed free path in which the tool can move and access the feature in a part to machine it. For each feature to be machined, the TAD is to be identified first. TAD of a feature is one of the most important factors in setup planning. A prismatic part can have six TADs in +X, −X, +Y, −Y, +Z, and −Z directions, as shown in Fig. 1.7a. A feature may have a single TAD or multiple TADs. Figure 1.7b shows different features with their possible TADs. A group of features are machined in a setup without repositioning the part. For formation of setups, features are clustered into different groups. TAD of a feature is the primary consideration for grouping features. Each group is assigned into different setups based on the TAD of the group. Features with a common TAD are generally grouped into the same setup. The features having multiple TAD are assigned a single TAD based on its tolerance relation with other features. Total number of setups depends on the machine capability in

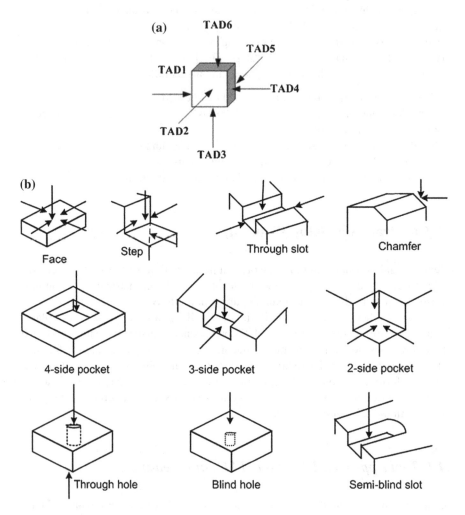

Fig. 1.7 Tool approach direction (TAD). **a** Six available TAD for a prismatic part. **b** Different features with their TADs

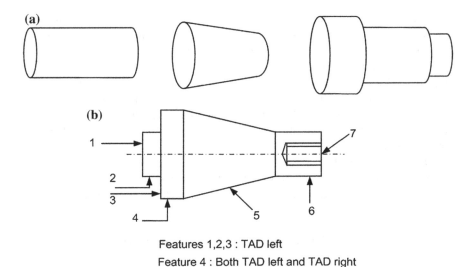

Features 1,2,3 : TAD left

Feature 4 : Both TAD left and TAD right

Features 5,6,7 : TAD right

Fig. 1.8 **a** Rotational parts. **b** Different features with their TAD for a rotational part

respect of feature access direction for machining. For a conventional milling or drilling machine, there can be maximum six numbers of setups considering the six TADs of a prismatic part.

In case of machining of rotational parts, the features can have only two possible TADs—from the right and from the left as shown in Fig. 1.8b. The features 1, 2, 3 have to be machined from left as they have TAD from the left. Similarly the features 5, 6 and 7 are to be machined from right since the TAD of the above features is from the right. Feature 4 can be machined either from the left or from right. As a result, only two setups setup-left and setup-right are possible for machining of rotational parts, whereas more than two setups are possible for machining prismatic parts. Features are clustered into two groups, TAD left and TAD right based on their TAD. Feature group with TAD left is assigned to setup-left and feature group with TAD right is assigned to setup-right.

1.4.2 Tolerance Requirements

Attaining the specified design tolerances is a key factor for the quality as well as the functionality of a machined part. The ultimate goal of setup planning is to achieve desired part quality at the lowest possible cost and time. To attain critical tolerance relationship between two features of a part, priority wise, the following setup methods are to be used [16]:

Setup method 1: In this method, two features are machined in the same setup with the same datum so that setup errors are eliminated. Tightest tolerance features are to be preferably machined in the same setup. In setup method 1, the tolerance relationship is only influenced by the machine tool motion error.

Setup method 2: In this method, one feature is used as datum for machining the other feature. This method is less accurate than setup method 1 as the tolerance relationship is influenced by both the machine tool motion error and setup error.

Setup method 3: Here, an intermediate datum is used to machine the two features in two different setups.

Setup method 3 is the least preferred method as a tolerance chain is formed for the dimensions obtained using this method. Tolerance will stack up and the resulting dimensions are less accurate compared to setup methods 1 and setup method 2.

1.4.3 Feature Interaction and Precedence Constraints

Among the features comprising a part, certain feature interactions take place. In many cases, two or more machining features interact among them. Feature interaction leads to precedence relations in the machining sequence of the features. An interaction between features occurs when machining of one feature affects the subsequent machining of another feature. There may be area feature interaction where two features share a common face. In volumetric feature interaction, there is common volume to be removed. Parent-child type of feature interaction occurs where the child feature is embedded in the parent feature. Different precedence relations are obtained due to area/volume feature interactions, tolerance relations, feature accessibility, tool/fixture interaction, datum/reference/locating requirements, etc. Different strategies are applied to machine interacting features satisfying the precedence relations. A precedence relation between two features F1 and F2, denoted as F1 → F2, implies that F2 cannot be machined until the machining of F1 is complete.

1.4.4 Fixturing Constraints

Each setup has to fulfil some fixturing constraints to be fixturable and feasible at the same time. Locating accuracy, complete restraint of the workpiece, minimum deformation at the fixture-workpiece interface, fixturing stability, non-interference of the tool and fixture are some of the fixturing related constraint.

1.4.5 Datum and Reference Constraints

The datum and reference requirements lead to the constraints that datum/reference features are to be machined prior to the related feature. As other features are located and dimensioned with respect to datum/reference features, these features are to be machined first.

1.4.6 Constraints of Good Manufacturing Practice

There are some rules of thumb evolving from decades of experience which are practised in the industry. These are considered as good manufacturing practice. For example, in case of drilling of two concentric holes, a hole of smaller diameter is drilled prior to a hole of larger diameter. Similarly, the hole of longer depth is drilled prior to the hole of shorter depth if they are concentric. Drilling a hole first and then reaming for enlargement is preferred. Another example of good manufacturing practice is using the highest area face for primary datum for stability.

1.5 Approaches of Process Planning

Since setup planning constitutes the core of process planning, it is very important to learn about the various approaches of computer aided process planning and setup planning. The approaches of computer aided process planning (CAPP) can be broadly divided into two categories; variant CAPP and generative CAPP. The following sections briefly present the variant and generative CAPP approaches.

1.5.1 Variant CAPP

The variant approach was the first approach used by the CAPP developers. Variant CAPP has evolved from the traditional manual process planning method. The variant CAPP is implemented based on group technology (GT) and parts classification and coding system. GT uses similarities between parts to classify them into part families. The different steps in variant process planning are:

Part family formation: The first step in implementing variant CAPP is to adopt an appropriate classification and coding system for the entire range of parts produced in the shop. The parts are coded classified into part families according to their geometric similarities and manufacturing characteristics. In the context of machining process planning, a part family consists of a set of parts that have similar machining requirements. All the existing parts are coded following the adopted scheme for coding. Each part family is represented by a family matrix and stored in the database with a family number. Example of some commercially available GT coding systems are Opitz coding, KK-3 coding, MICLASS coding, DCLASS coding, etc.

Standard process plan preparation: The next step is to prepare a process plan, also called the standard process plan for each part family. The standard process plans are then stored in a database and indexed by part family matrices. After completion of the above steps, the variant CAPP system is ready for use.

Preparatory phase   Process planning phase

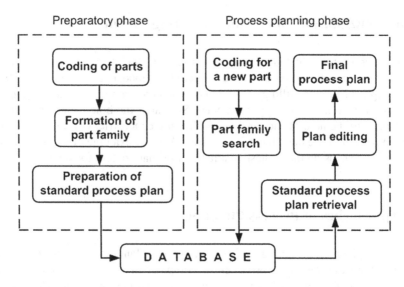

Fig. 1.9 A variant CAPP approach

Process planning for a new part: To obtain a process plan for a new part, first the part is coded. The code of the part is used to search the database of part family matrix to retrieve a standard process plan for a similar part. Therefore variant CAPP is also known as retrieval CAPP.

Plan editing: The standard process plan is examined to determine if any modifications are necessary. The process planner edits the process plan according to the requirements of the new part to create a variant of the existing process plan. If a standard process plan does not exist for the given code, then a similar code for which a standard plan exists may be searched. The new plan can be put into the part family matrix for future reference.

Implementation: If the new part belongs to an existing part family, the standard plan for the part is retrieved, modified and implemented. On the other hand, if the new part does not belong to an existing part family, a new process plan is developed manually and stored in the part family matrix. The schematic diagram of a variant CAPP approach is shown in Fig. 1.9.

Variant approach has obvious disadvantages. Its limitations are that it is restricted to similar parts previously planned and experienced process planners are required to modify the standard plan. Quality of the process plans depends on the knowledge and experience of the process planner. It is unable to automatically generate process plans. Moreover, the cost involved in creating and maintaining database for the part families is high. Due to these problems, variant approach is normally used when a well-defined part family structure exists and the new part closely conforms to the characteristics of the existing part families. One of the earliest variant CAPP approaches CAPP, CAM-I can be traced back to 1976 [26]. Some other variant process planning examples are Computer-Aided Process Planning (CAPP) [38], Technostructure of Machining (TOM) [29], and Group Technology Works (GTWORKS) [21].

1.5.2 Generative CAPP

The generative CAPP approach represents an alternative approach to automated process planning. The generative CAPP approach automatically creates process plans for a new part from scratch without referring to existing plans. Manufacturing knowledge is encoded into efficient software and automatic generation of process plan for a new part is possible using manufacturing knowledge and manufacturing information database. By applying decision making mechanism, a process planner's decision making logic is imitated. The decision making mechanism can be procedural algorithms, decision trees, decision tables and production rules. A major advantage of generative CAPP systems over variant systems is that they can provide process plan for a part for which no variant of the part exists which can be retrieved and modified. Another advantage of the generative approach is the generation of more consistent plans. Some examples of generative process planning systems are Expert Computer-Aided Process-Planning (EXCAP) [9], Process Planner (PROPLAN) [35], Cutting Technology (CUTTECH) [3], Semi Intelligent Process Planning (SIPP) [31], Hierarchical and Intelligent Manufacturing Automated Process Planner (HI-MAPP) [4], and Quick Turnaround Cell (QTC) [22]. The main components of a generative CAPP system are shown in Fig. 1.10. They are briefly described in the following paragraphs.

Part representation database: Part representation is one of the major factors of generative CAPP. The part representation database comprises the input information to a generative CAPP system. It contains part information including shape,

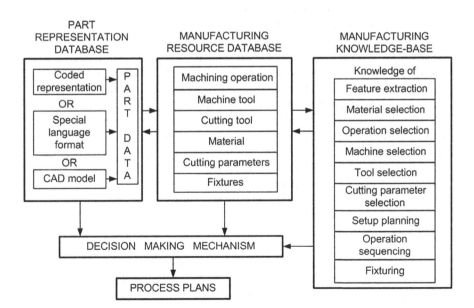

Fig. 1.10 A generative CAPP approach

geometric features, dimensions, tolerances and surface condition. Different methods have been used for representing the part for generative CAPP. They include GT codes, special descriptive languages and feature based CAD models.

Although GT codes are commonly used in variant CAPP systems, they have been used as well for part description in some generative CAPP systems [7, 18, 24, 25, 30, 42]. However, the code based part representation is not suited for automated process planning system as coding is a manual process. It is both time consuming and prone to error. Another major disadvantage of GT based coding systems is the cost involved in creating and maintaining databases for the part families.

Special descriptive languages have been used for part description in generative CAPP systems such as AUTAP-NC [12] and GARI [10]. Basic part geometry is translated into higher level format that can be used by the process planning system. Expert system based CAPP approaches generally use special descriptive language for input part description. This form of part representation is also not suitable for automated process planning as it requires a lot of effort by the user who has to manually prepare the input data.

Rapid development of feature based 3-D CAD models has made them the preferred choice of the CAPP developers for part representation. Widespread use of solid modelling has led to design by features approach where the designer designs a part in terms of its features. The CAD model contains detail information about a part and it provides the necessary information for process planning functions. The use of CAD models for input part representation in generative CAPP systems can eliminate the human effort required in case of GT codes or special descriptive languages.

Manufacturing resource database: This module contains information on the manufacturing resources needed for production. It may include information on material and machinability, machine tool, cutting tool, machining processes and their process capabilities, available process parameters, fixtures, material handling system, etc.

Manufacturing knowledge-base: Knowledge-base contains the manufacturing knowledge that is commonly used by human process planners for machining a component. For example, the knowledge of matching the machining requirements of a part to the process capabilities of machining processes is stored in the manufacturing knowledge-base. Knowledge-base contains knowledge for feature extraction, machining operation selection, machine and tool selection, operation sequencing, selection of cutting conditions, jig and fixture selection, etc. The knowledge is elicited by the CAPP developer from various experts, books and manuals. The collected knowledge is represented and stored in the knowledge-base in various forms such as production rules, semantic networks, conceptual graphs, frames, object oriented schemes and Petri nets [33].

Decision making mechanism: The decision making mechanism of generative process planning includes and realizes the decision making logic used by the process planner to make decisions of process planning. The decisions may be on selection of material, selection of machining operations, machine tools and cutting tools, setup planning, selection of jigs and fixtures, determination of cutting conditions and so on. It contains an inference engine that links the knowledge in the knowledge-base and the data in the part and manufacturing resource database to form a line of reasoning to take decision. Different methods have been used to represent the decision making logic in generative CAPP. Traditional methods such as decision trees, decision tables, algorithms and graph theory have been used for decision making in CAPP. An example of the application of decision tree in generative CAPP is Expert Computer-Aided Process-Planning (EXCAP) [9]. M-GEPPS is another CAPP system which uses decision tree for process planning of rotational parts [42]. Cutting Technology (CUTTECH) [3], Hierarchical and Intelligent Manufacturing Automated Process Planner (HI-MAPP) [4] are two CAPP systems which use decision tables for decision making. However, owing to several disadvantages of the traditional methods, there is a gradual shift towards the use of artificial intelligence (AI) based methods for decision making.

1.6 Setup Planning for Prismatic Parts

In the last three decades, extensive research work is carried out in the area of setup planning for machining a component. Considering the part geometry, these works can be categorized into setup planning for prismatic parts and setup planning for cylindrical or rotational parts. Although the nature of the problem appears to be similar for both the cases, different approaches are to be adopted for them. The main differences in setup planning for prismatic and rotational parts lie in the type of features, tool approach direction (TAD) of a feature, selection of locating and clamping features and type of fixtures.

A part can be called prismatic if it has two opposite parallel faces as polygons and the other faces are rectangles. Figure 1.11 shows some stocks of prismatic parts. It may contain different types of features, e.g. flat face, step, slot, pocket, chamfer, hole, keyway, etc. Tool approach direction (TAD) of a feature is one of the most important factors in setup planning. TAD of a feature is discussed in detail in Sect. 1.4.1. For each feature to be machined, the TAD is to be identified first. TAD of a prismatic part is shown in Fig. 1.7a. A feature may have a single TAD or multiple TAD. Figure 1.7b shows different features of a prismatic part with their possible TAD. For formation of setups, features are clustered into different groups. TAD of a feature is the primary consideration for grouping features.

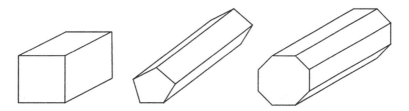

Fig. 1.11 Prismatic parts

Each group is assigned into different setups based on the TAD of the group. Features with a common TAD are generally grouped into the same setup. Total number of setups depends on the machine capability in respect of feature access direction for machining. For a conventional milling or drilling machine, there can be maximum six numbers of setups considering the tool access directions of a prismatic part. However, it is possible to machine five faces of a cubic workpiece in a single setup in a modern machining center (MC) equipped with rotary index table and automatic tool changer (ATC). Most of the machining centers contain simultaneously controlled three Cartesian axes X, Y, and Z. Therefore it is possible to machine all the features of a part in these machines in less number of setups compared to conventional machines.

The geometry of a part plays a key role in the selection of the type of fixtures to be used for machining. A fixture is a single device or a combination of components for locating, supporting and holding a part during machining. Different types of fixtures are used for fixturing prismatic parts. They are custom made fixtures, vices, modular fixtures, etc. Custom made fixtures are costly, inflexible and time consuming in designing. A vice has a stationary jaw for part locating and a movable jaw for part clamping. It can be employed for both vertical and horizontal machines. In spite of their ease of use, vices are restricted by their sizes. A modular fixture has a flat base plate, locators, clamps and supports for fixturing the part. Modular components can be assembled like building blocks on the base plate. They can be re-used and adapted for different parts. Modular fixtures are capable of handling a wide variety of part sizes and shapes. Moreover, they are cost and time efficient. These qualities have made modular fixtures the most preferred choice for machining of prismatic parts.

Different features of a prismatic part are used for locating and clamping. By location, the position and orientation of a part is established relative to the machine tool. The part is held in the required position with clamps. Generally three types of features are used for location in case of prismatic parts; planar surfaces, holes and external profiles. There are mainly two locating methods for prismatic parts; 3-plane locating (3–2–1) and 1-plane and 2-hole locating. For a prismatic part, the planar surfaces of the part can be used conveniently for 3–2–1 locating and this locating method is mostly used for prismatic parts. Normally, the largest surfaces opposite to the locating surfaces are used for clamping. The position of the clamp should be so chosen that there is no interference between the clamping elements and the cutting tool. For details of different locating and clamping methods, the reader is directed to Joshi [20].

1.7 Setup Planning for Cylindrical Parts

A part can be called cylindrical or rotational if it has an external cylindrical surface with circular cross section. Figure 1.8a shows some rotational parts. It may contain different types of features, e.g. cylindrical face, external planar face with circular area, hole, groove, slot, external taper, external thread holes, etc. In case of machining of rotational parts, the features can have only two possible TAD; from the right and from the left as shown in Fig. 1.8b. The features 1, 2, 3 have to be machined from left as they have TAD from the left. Similarly the features 5, 6 and 7 are to be machined from right since the TAD of the above features is from the right. Feature 4 can be machined either from the left or from right. As a result, only two setups, setup-left and setup-right, are possible for machining of rotational parts, whereas more than two setups are possible for machining prismatic parts. Setups are formed in the similar manner as in case of prismatic parts. Features are clustered into two groups, TAD left and TAD right based on their TAD. Feature group with TAD left is assigned to setup-left and feature group with TAD right is assigned to setup-right.

Different fixtures used for rotational parts are chucks, face plates, collets, etc. Mainly chuck-type fixtures are used for rotational parts. It employs radially adjustable jaws to define the axis of rotational part. Generally for a rotational part, end faces are selected as locating features and external cylindrical faces are selected as clamping features.

1.8 Flexibility in Setup Planning

In today's manufacturing scenario, importance is shifted from isolated manufacturing system to flexible manufacturing system in geographically distributed manufacturing environment. In the emerging trend of agile and virtual manufacturing, a part can be designed, process planned and manufactured in different manufacturing sites across the globe. Java and Web technologies provide a common platform for collaborative design and manufacturing enabling transfer of information between various manufacturing systems. The recent developments in internet technologies can be utilized to integrate different manufacturing systems located at different sites to enable exchange of information among them. Web-based manufacturing systems have been developed to facilitate sharing of production knowledge through the internet. Java and Web technologies coupled with eXtensible Modeling Language (XML) file format provide means for the transfer of information between various manufacturing systems. Researchers have developed setup planning approaches which can be integrated with the Internet [15, 41]. The proposed systems have client/server architecture comprising an information server, a database server, and a number of setup planning clients. The use of Java and XML adds flexibility to the systems and operable under different platforms.

Another important consideration in process planning is adaptability. Adaptation to the changing scenario is a crucial factor in this era of lean and agile manufacturing. There is continuous improvements and redesigning of products in the competitive market. Therefore the setup plans/process plans have to be flexible and ready to adapt to the changes in design as well as manufacturing environment. Setup plans developed prior to actual production may become infeasible during actual production time due to changes in the conditions on the shop floor. Traditional software systems for automating setup planning are static in nature and they do not respond to the changes in the situation. To make a setup plan adaptable to the actual manufacturing situation, the shop floor status is to be continuously monitored. For any change in the status, the system should be able to revise the information stored earlier and provide the current status information. This may be applicable to change in the shop floor capacity, machine breakdown, tool breakdown, change in the routing, etc. Different approaches to develop setup plans adaptable to changing scenario are found in the literature which are responsive to the user's changing needs [5, 13, 28, 39, 40]. Azab and ElMaraghy [1, 2] contributed towards a new concept of reconfigurable process planning (RPP) which is a semi-generative process planning approach suitable for agile and reconfigurable manufacturing system (RMS) environment. Reconfigurable machine tools play an important role in dynamic and adaptable manufacturing systems.

Integration among different functional areas within a manufacturing industry is essential for improvements in quality, efficiency, cost and time. Each of design, process planning and production planning, manufacturing, quality control and other support functions act as a part of a unified system rather than a stand alone system. Therefore, the need arises to integrate all these phases to allow a path from initial concept to a finished product. Automation is gaining prime importance in modern manufacturing industries to fulfill the need for improved productivity and quality. Many sophisticated CAD and CAM systems are available commercially. However, equivalent commercial automatic setup planning/process planning systems are sparse. An important issue in integration and automation is the incompatibility of the equipments and softwares. Development of proper interface standards is necessary for exchange of information among design, process planning and manufacturing stages.

Proper interpretation of the part design and extracting accurate part information is a major factor in process planning/setup planning. Design information of the part constitutes an important input to process planning for machining a part. The part representation database of a CAD system comprises information of the part including features of the part, part dimensions, shape, tolerances, surface finish, etc. CAPP has to extract manufacturing information such as machining features and precision specifications of the part including surface roughness, and dimensional and geometric tolerances to machine the part. However, CAD/CAPP/CAM systems normally have different product data descriptions. CAD information about a product is usually geometry-based and CAPP and CAM are feature-based leading to incompatibility in practical implementation. The CAD data can not be used directly on a machine to cut a component. CAD focuses on part specific geometry while process planners and operators are more concerned with process-specific features and their accuracies. Different methods have been used by researchers to address the incompatibility issue in CAPP. Interfacing CAD system with a Group Technology (GT) coding system or an interface standard like Standard for the Exchange of Product Model Data (STEP) or Initial Graphics Exchange Standards (IGES) is a common practice for generating feature information for process planning. STEP compliant neutral file can be used to connect dissimilar CAD packages to CAPP. Instead of coding schemes having a rigid digit length structure, flexible digit length can be used so that it becomes possible to include all the detail of the features in the codes. The benefit of these types of approaches is that the product data from various CAD systems can be interconnected and automatically coded for multiple manufacturing purposes.

Addressing the above mentioned issues leads to the need of automated, flexible and web based setup planning/process planning systems. A brief discussion on commonly used interface standards in manufacturing is presented in the following section. Two popular standards are IGES and STEP.

1.8.1 Exchange of Product Data

Integration of different functional areas of a manufacturing system is of utmost importance for bridging the gap between the initial concepts of a product to its finished stage. One of the major difficulties in integrating CAD-CAPP-CAM is the communication gap due to incompatibility of equipment and software. Each CAD/CAPP/CAM system has its unique format of storing data and special translator programs are needed for exchanging data with other CAD/CAPP/CAM systems. The inconvenience of developing special translator programs led to the idea of a standard generic data exchange format that can be used by all CAD-CAPP-CAM users for effective communication of product data. Over the last three decades, a number of data exchange standards have been developed, e.g. Initial Graphics Exchange Standards (IGES), Standard for the Exchange of Product Model Data (STEP), Data eXchange Format (DXF), ACIS, etc. However, IGES and STEP are the most widely accepted standards used by the CAD-CAPP-CAM users.

1.8.1.1 IGES

Initial Graphics Exchange Standards (IGES) was the first common platform data exchange standard developed by major US CAD vendors with support from US National Bureau of standards in 1979. It was adopted as ANSI standard in 1981 and later as an ISO international standard. Since its inception, it has gone through many modifications and the current version supports 2-D, 3-D and surface modelling data exchange between CAD-CAPP-CAM systems. IGES offers a neutral file format where the modelling database of a given product can be described as an IGES model. The purpose of this IGES model is to provide an intermediate data file which can be shared and interpreted by different CAD-CAPP-CAM systems. CAD systems supporting IGES have to add processor software to translate data files from originating software to IGES neutral format. An IGES file contains the following sections in order:

Flag section: optional section that precedes the **start** section

Start section: added manually by the initiator of the IGES file that contains introductory information about the source CAD/CAM system and a brief description of the product

Global section: contains information about the preprocessor and postprocessor for interpretation of the IGES file, e.g. software ID, sender and receiver's identifier, IGES processor version, necessary parameters to translate the file, precision of the numbers, resolution and coordinate values, model space scale, etc.

Directory section: generated by the pre-processor and contains a list of all the entities (points, lines, arcs, etc.) in the IGES file with the attributes associated with them. For example, a line entity with its associated attributes such as line type, font, colour, weight are listed

Parameter data section: contains parameters associated with each entity in the IGES file, e.g. six coordinates of the two end points of a line entity, number of points used for construction, annotation text if any and so on. Number of parameters used varies from entity to entity and the first parameter identifies the entity type

Termination section: signifies the end of the IGES file and contains the total number of records of the previous sections.

1.8.1.2 STEP

Standard for the Exchange of Product Model Data (STEP) is an ISO international standard which was initiated in 1984 in search of a robust and more generic data exchange standard than IGES. It is the result of European and US collaborative research venture which has received worldwide acceptance and popularity. STEP tries to overcome some of the limitations of IGES. The aim of STEP is to define a standard file format that contains all the information of a product right from the design stage through manufacturing, quality control, product functionality, and testing. The data transferred through STEP includes part geometry with all the detail, analysis, manufacturing steps with process planning, quality assurance and testing procedures. STEP can be used for multiple application domains, e.g. mechanical, electrical, electronics engineering areas. STEP files are interpretable by computer that makes it suitable for automation purpose. STEP contains the following sections:

Implementation section: contains the implementation techniques. A specially developed information modelling language EXPRESS is used to describe the product model and the file format that stores it. EXPRESS maps the model data to the original CAD file. It follows an object oriented approach, e.g. a circle is considered an object and defined by its center coordinates and radius

Resource information section: contains information on geometry, topology and structure of the product, tolerances, visual presentation, materials, process structure and properties and provides to the next section

Application protocol (AP): contains information for specific application domains such as mechanical, electrical, electronics, automotive, aerospace, etc. A particular STEP format can be used for a particular industrial application, e.g. AP207 is used for sheet metal die planning and design and AP210 is used for electronic assembly design

Abstract test suites: contains test suites for the application protocols provided for different industrial applications

Application interpreted construct: contains information on different model entity constructs (edge-based, shell-based, and geometrically bounded wireframes, topologically and geometrically bounded surfaces, etc.) and specific modelling approaches (constructive solid geometry, wireframe modelling, surface modelling, etc.).

STEP provides a better alternative to IGES and it is undergoing continuous evolution process for the better. Most of the current CAD systems are incorporating STEP. One of the many achievements of STEP is the integration with eXtensible Markup Language (XML) which is a very convenient tool for online data access through the web. There are three main reasons for implementing STEP [44]:

- Data Exchange—Exchange product data with consumers and suppliers.
- Data Sharing—Store product data in a standard database for use by external and internal supply chains.
- Internet Collaboration—Product data can be easily accessed via Internet.

1.9 Setup Planning for Green Manufacturing

Environmental consideration in manufacturing is an important issue in the present manufacturing scenario. Due to the increasing industrial regulations of the government and growing consumer preference for green products, manufacturers have begun to explore means of reducing energy requirements and impact on the environment through improved process design and planning. For this, focus is given on resource consumption and manufacturing planning. The objective of the process/setup planning for green manufacturing is to improve the green quality by reducing resource consumption and environmental impacts of the machining process. It has become an essential and fast developing research area in manufacturing. Process planning for green manufacturing encompasses evaluation of the resource consumption and energy consumption, reducing environmental impact of the manufacturing processes, reducing noise, health hazards and waste emissions and enhancing security.

Literature review on process planning and setup planning reveals that there is some awareness for green engineering and green manufacturing in the last two decades. There have been several efforts to develop process plans for green manufacturing [8, 11, 17, 19, 23, 27, 32, 34, 36, 45, 47]. The objective of green process planning is to optimize the raw material consumption, secondary material (e.g. coolant) consumption, energy consumption, and environmental impacts to make the manufacturing processes eco-friendly. Exploring for the improved ways of green process planning can give a new direction to the research on environment friendly manufacturing.

1.10 Conclusion

In this chapter the fundamentals of machining process planning and setup planning are presented. The need and challenges of automated process planning/setup planning are discussed and the main constraints of setup planning are enumerated. Some important concepts for setup planning are presented in detail. The variant and generative approaches of process planning are discussed with detail on part information input format, decision making strategy. The main differences in the method of setup planning for prismatic and rotational parts are presented. Finally the importance of green machining is discussed.

References

1. Azab A, ElMaraghy HA (2007) Mathematical modeling for reconfigurable process planning. Ann CIRP 56:467–472
2. Azab A, ElMaraghy HA (2007) Sequential process planning: a hybrid optimal macro-level approach. J Manuf Syst 26:147–160
3. Barkocy BE, Zdeblick WJ (1984) A knowledge based system for machining operation planning. In: Proceeding of AUTOFACT, vol 6, Arlington, Texas, pp 2.11–2.25
4. Berenji RH, Khoshnevis B (1986) Use of artificial intelligence in automated process planning. J Comput Mech Eng 5:47–55
5. Cai N, Wang L, Feng HY (2008) Adaptive setup planning of prismatic parts for machine tools with varying configurations. Int J Prod Res 46:571–594
6. Chang TC, Wysk RA, Wang HP (2011) Computer-aided manufacturing. Dorling Kindersley (India) Pvt. Ltd, New Delhi
7. Chitta AK, Shankar K, Jain VK (1990) A decision support system for process planning. Comput Ind 14:307–318
8. Dahmus JB, Gutowski TG (2004) An environmental analysis of machining. In: Proceedings of ASME international mechanical engineering congress and RD&D Expo, 13–19 Nov, California, USA
9. Davis BJ, Darbyshire IL (1984) The use of expert systems in process planning. Ann CIRP 33:303–306
10. Descotte Y, Latombe JC (1981) GARI: a problem solver that plans how to machine mechanical parts. In: Proceedings of IJCAI-7, Vancouver, Canada, pp 766–772
11. Diaz N, Redelsheimer E, Dornfeld D (2011) Energy consumption characterization and reduction strategies for milling machine tool use. In: Proceedings of the 18th CIRP international conference on life cycle engineering, 2–4 May, Germany, pp 263–267
12. Eversheim PJ, Holz B (1982) Computer aided programming of NC machine tools by using the system AUTAP-NC. Ann CIRP 31:323–327
13. Feng SC, Zhang C (1998) A modular architecture for rapid development of CAPP systems for agile manufacturing. IIE Trans 30:893–903

14. Gao S, Shah JJ (1998) Automatic recognition of interacting machining features based on minimal condition subgraph. Comput Aided Des 30:727–739
15. Gaoliang P, Wenjian L, Xutang Z (2005) An internet-based system for setup planning in machining operations. In: Proceedings of the 10th IEEE international conference on engineering of complex computer systems, Shanghai, China, pp 245–251
16. Huang SH, Liu Q (2003) Rigorous application of tolerance analysis in setup planning. Int J Adv Manuf Technol 3:196–207
17. Jadhav PV, Bilgi DS (2008) Process planning support system for green manufacturing and its applications. In: Proceedings of Indo-Italian international conference on green and clean environment. MAEER's MIT College of Engineering, Pune
18. Jiang B, Baines K, Zockel W (1997) A new coding scheme for optimization of milling operations for utilization by a generative expert CAPP system. J Mater Process Technol 63:163–168
19. Jiang ZG, Zhang H, Xiao M (2008) Web-based process database support system for green manufacturing. Appl Mech Mater 10–12:94–98
20. Joshi PS (2000) Jigs and fixtures, 2nd edn. Tata McGraw-Hill Publishing Company Limited, New Delhi
21. Joshi SB, Hoberecht WC, Lee J, Wysk RA, Barrick DC (1994) Design, development and implementation of an integrated group technology and computer aided process planning system. IIE Trans 26:2–18
22. Kanumury M, Chang TC (1991) Process planning in an automated manufacturing environment. Int J Manuf Syst 10:67–78
23. Krishnan N, Sheng PS (2000) Environmental versus conventional planning for machined components. CIRP Ann Manuf Technol 49:363–366
24. Lau H, Jiang B (1998) A generic integrated system from CAD to CAPP: a neutral file-cum-GT approach. Comput Integr Manuf Syst 11:67–75
25. Lau HCW, Lee CKM, Jiang B, Hui IK, Pun KF (2005) Development of a computer-integrated system to support CAD to CAPP. Int J Adv Manuf Technol 26:1032–1042
26. Link CH (1976) CAPP, CAM-I automated process planning system. In: Proceedings of the 1976 NC conference, CAM-I Inc., Arlington, Texas
27. Liu F, Yan J, Cao H, He Y (2005) Investigations and practices on green manufacturing in machining systems. J Cent South Univ Technol 12:18–24
28. Masood A, Srihari K (1993) RDCAPP: a real-time dynamic CAPP system for an FMS. Int J Adv Manuf Technol 8:358–370
29. Matsushima K, Okada N, Sata T (1982) The integration of CAD and CAM by application of artificial intelligence techniques. Ann CIRP 31:329–332
30. Nadir Y, Chaabane M, Marty C (1993) PROCODE-automated coding system in group technology for rotational parts. Comput Ind 23:39–47
31. Nau DS, Chang TC (1985) A knowledge based approach to generative process planning. In: Symposium of computer-aided intelligent process planning, ASME winter meeting, Miami, Florida
32. Neugebauer RA, Drossel WA, Wertheim RB, Hochmuth CA, Dix M (2012) Resource and energy efficiency in machining using high-performance and hybrid processes. In: Proceedings of 5th CIRP conference on high performance cutting, Germany, pp 3–16
33. Nikolopoulos C (1997) Expert systems: introduction to first and second generation and hybrid knowledge based systems. Marcel Dekker Inc., New York
34. Park C, Kwon K, Kim W, Min B, Park S, In-Ha Sung I, Yoon YS, Lee K, Lee J, Seok J (2009) Energy consumption reduction technology in manufacturing—a selective review of policies, standards, and research. Int J Precis Eng Manuf 10:151–173
35. Philips RH, Zhou XD, Mouleeswaran CB (1984) An artificial intelligence approach to integrating CAD and CAM through generative process planning. In: Proceedings of ASME international computers in engineering, Las Vegas, US
36. Sheng P, Srinivasan M (1995) Multi-objective process planning in environmentally conscious manufacturing: a feature-based approach. CIRP Ann Manuf Technol 44:433–437

37. Tseng YJ, Joshi SB (1998) Recognition of interacting rotational and prismatic machining features from 3-D mill-turn parts. Int J Prod Res 36:3147–3165
38. Tulkoff J (1978) CAPP, CAM-I automated process planning system. In: Proceedings of 15th numerical control society annual meeting and technical conference, Chicago
39. Usher JM, Farnandes KJ (1996) A two-phased approach to dynamic process planning. Comput Ind Eng 31:173–176
40. Wang L, Feng HY, Cai N (2003) Architecture design for distributed process planning. J Manuf Syst 22:99–115
41. Wenjian L, Gaoliang P (2005) An internet-enabled setup planning system. In: Proceedings of the third international conference on information technology and applications, Sydney, Australia, pp 89–92
42. Wysk RA, Chang TC, Wang HP (1988) Computer integrated manufacturing software and student manual. Delmar Publishers, New York
43. Xu N, Huang SH, Rong YK (2007) Automatic setup planning: current state-of-the-art and future perspective. Int J Manuf Technol Manage 11:193–207
44. Xu XW, He Q (2004) Striving for a total integration of CAD, CAPP, CAM and CNC. Robot Comput Integr Manuf 20:101–109
45. Yan HE, Fei L, Huajun C, Hua Z (2007) Process planning support system for green manufacturing and its application. Front Mech Eng China 2:104–109
46. Yan X, Yamazaki K, Liu J (2000) Recognition of machining features and feature topologies from NC programs. Comput Aided Des 32:605–616
47. Yeo SH, New AK (1999) A method for green process planning in electric discharge machining. Int J Adv Manuf Technol 15:287–291

Chapter 2
Different Phases of Setup Planning

Abstract In this chapter different phases of setup planning task are discussed in detail. Setup planning mainly comprises of feature grouping, setup formation, datum selection, machining operation sequencing, and setup sequencing. The main criteria for feature grouping and setup formation are tool approach direction and tolerance relation among the features. Datum selection primarily depends on area of a feature, its orientation, surface quality and its tolerance relations with other features. Machining operation sequencing and setup sequencing is done based on feature precedence relations.

Keywords Features · Datum · Setups · Feature precedence relation · Operation sequencing

2.1 Introduction

Setup planning is an important intermediate phase of process planning. Output of a setup planning system gives the necessary instructions for setting up parts for machining. Setup planning consists of various phases such as feature grouping, setup formation, datum selection, machining operation sequencing, and setup sequencing. It takes information on features of a part, machining operations, machine tools and cutting tools as inputs from part representation database and manufacturing resource database. The part representation database comprises the information of the part including features of the part, part dimensions, shape, tolerances, surface finish, etc. Similarly, manufacturing resource database comprises information of machining operations, machine tools, cutting tools, materials, etc. Based on these inputs, manufacturing knowledge, and constraints in setup planning (discussed in Sect. 1.4), setup planning is performed. Different phases of setup planning are discussed in detail in the following sections.

© The Author(s) 2015
M. Hazarika and U.S. Dixit, *Setup Planning for Machining*,
SpringerBriefs in Manufacturing and Surface Engineering,
DOI 10.1007/978-3-319-13320-1_2

2.2 Feature Grouping

A part to be machined contains a number of machining features. The machining features represent the geometry of a part. A raw stock is converted to a finished part after machining these features on it. A group of features are machined in a setup without repositioning the part. Features to be machined in a particular setup are grouped together and machined in a particular machining sequence. Machining of the maximum number of features in the same setup ensures better tolerance achievement. The different features of the part are assigned to different setups based on several criteria such as tool approach direction (TAD) of the feature, tolerance requirements, precedence relations among the features, feature geometry, and feature interactions. Clustering of features and their machining operations into different groups is primarily done based on their TADs. For each feature to be machined, the TAD is to be identified first. A prismatic part can have six TADs and a rotational part can have two TADs as shown in Figs. 1.7 and 1.8 respectively. A feature may have a single TAD or multiple TADs. Another important criterion for feature grouping is tolerance relations among features. Normally, features with tight tolerance relations are assigned to the same setup. The following methodology is adopted for grouping of features for setup formation.

- Features with a common single TAD are grouped together to form a common TAD feature cluster. A common TAD feature cluster can be machined in the same setup.
- A feature having multiple TADs can be assigned to different TAD feature clusters and thus alternative machining sequences can be obtained for the same component. Alternatively, it can be assigned a single TAD based on its tolerance relations with other features. For example, if a multiple TAD feature (say a) has tolerance relation with only one feature (say b) having a single TAD common with a, then the feature a is assigned the TAD of b.
- If a multiple TAD feature (say a) has tolerance relation with more than one feature (say b and c) each having a single TAD, then the feature a is assigned the TAD of b or c, depending on whichever has tighter tolerance relationship with a.
- If a multiple TAD feature has no tolerance relationship with other features, it is assigned the TAD of a feature cluster where there are the maximum numbers of features. Machining of the maximum number of features in the same setup with the same datum will ensure better tolerance achievement and reduced machining time and cost.

To explain the method described above, the following example is taken. Figure 2.1 shows a component to be machined along with the detailed information on its features, dimensions, machining operations needed, TAD and tolerances among the features.

In Fig. 2.1, all the six faces (faces 1, 2, 9, 10, 11, and 12) of the prismatic block are initially rough machined and only faces 1 and 2 are considered as machining features. The through hole 8 has parallelism tolerance 0.15 mm with the blind hole 7

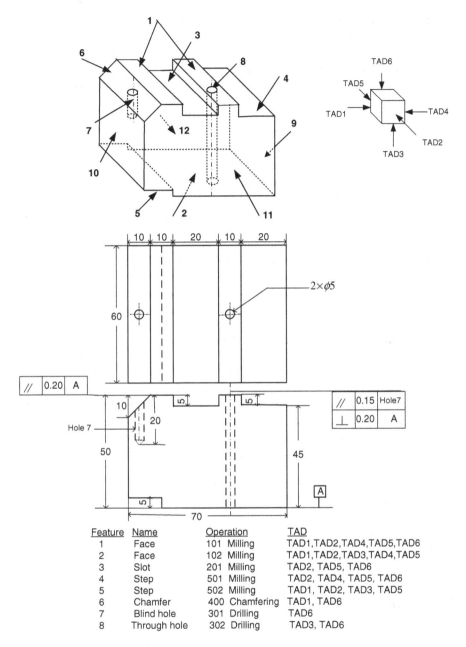

Feature	Name	Operation	TAD
1	Face	101 Milling	TAD1,TAD2,TAD4,TAD5,TAD6
2	Face	102 Milling	TAD1,TAD2,TAD3,TAD4,TAD5
3	Slot	201 Milling	TAD2, TAD5, TAD6
4	Step	501 Milling	TAD2, TAD4, TAD5, TAD6
5	Step	502 Milling	TAD1, TAD2, TAD3, TAD5
6	Chamfer	400 Chamfering	TAD1, TAD6
7	Blind hole	301 Drilling	TAD6
8	Through hole	302 Drilling	TAD3, TAD6

Fig. 2.1   A component with its features

and perpendicularity tolerance 0.20 mm with face 2, so it has a tighter tolerance relation with 7. Face 1 has parallelism tolerance 0.20 mm with face 2. Face 2 also has
positional tolerance relations with features 4, 5, and 6. Through hole 8 has two TADs

and it can be assigned TAD6 based on its tighter tolerance relation with feature 7. Features 1, 3, 4 and 6 have multiple TADs and they can be assigned to TAD6 feature cluster where there is the maximum number of features which will ensure better tolerance achievement and reduced machining time and cost. Similarly features 2 and 5 are assigned to TAD3 feature cluster. Thus, all the features can be incorporated into two different TAD feature clusters, viz. TAD6 and TAD3 feature cluster.

2.3 Setup Formation

After grouping of features based on TAD and tolerance relations, setups are formed. In each setup, a number of features are to be machined. For setup formation, different common TAD feature clusters are grouped together considering the machine capability. Total number of setups depends on the machine capability in respect of feature access direction for machining. For a conventional milling or drilling machine, there can be maximum six setups for machining prismatic parts considering their six TADs. Nowadays, various milling as well as drilling operations can be performed in a modern machining center (MC) equipped with rotary index table and automatic tool changer (ATC). Most of the machining centers contain simultaneously controlled three Cartesian axes X, Y, and Z. It is possible to machine five faces of a cubic component in these machines in a single setup. The five common TAD feature clusters (TAD1, TAD2, TAD4, TAD5 and TAD6 as shown in Fig. 1.7) can be grouped into one setup and the remaining common TAD feature cluster TAD3 can be assigned to the other setup. The component can be machined using only two setups compared to six setups of conventional machines.

For rotational parts, features and their machining operations for a given machine tool are clustered into two groups or two setups: (i) machining operations to be performed from the right and (ii) machining operations to be performed from the left. The proper decision is to be taken after considering the TADs and relative tolerance relationships among the features. Note that only two setups—setup-left and setup-right are possible for machining of rotational parts. For example, for the rotational part shown in Fig. 1.8b, features 1, 2, 3, and 4 can be assigned to setup-left and 5, 6, and 7 can be assigned to setup-right.

2.4 Datum Selection

In setup planning, selection of proper datum is essential for attaining the specified tolerances of the machined component. For creating reference for a component to be machined, datum is used. Once the features to be machined are grouped and setups are formed, datum for each setup is to be selected. Setup datum provides a definite and fixed position for machining the component. Datum planes and datum features are discussed in Sect. 1.2.4. Generally datum features rest on datum planes. The imaginary plane on which a component lies during machining is called the primary datum plane. The actual feature of the component that lies on the primary datum plane is called the primary datum. For prismatic components, primary datum is normally a face of the component, resting on which the features in a setup undergo machining. However, a datum feature may be a face, an axis, a curve or a point. In case of rotational components, both holes and surfaces can be used as datum features. Datum selection is the task of identifying the potential features which can serve as primary, secondary and tertiary datum for each setup. Features sharing common TAD and datum are naturally grouped into one setup.

Selection of the proper datum is one of the most challenging tasks in setup planning [8]. The approaches found in the literature for selection of datum are diversified in terms of criteria considered, such as total area of a face, its orientation, tolerance relation with other features, stability it provides, and symmetry and intricacy of a face. Large and maximum area face has been the most widely used criterion for selecting the primary datum for machining [3, 14]. However, surface area is not the only consideration for selecting datum. For proper location, the surface quality of datum is also important. It is well recognized that surface finish is one of the criteria for assessing the suitability of a face to be selected as datum [2, 9, 13, 15, 18]. Usually, the datum surfaces are the machined surfaces. However, it is to be noted that Hazarika et al. [8] observed that under some circumstances, excessively smooth surface as datum may produce more manufacturing errors compared to a rough surface datum. Many researchers consider tolerance relations among features as the prime criteria for selecting datum [1, 6, 7, 11, 17]. Selection of proper datum is very important for tolerance requirements and functionality of the part. To select datum for a setup in case of a prismatic part, first all the faces of the part are identified. The faces having an orientation different from the faces being machined in that setup are sorted out. Then, they are assessed for suitability as datum based on the above mentioned criteria.

In case of rotational parts, the surface which has an orientation different from the surfaces being machined (for rotational parts, two orientations: orientation from the left and that from the right is possible) is selected as datum. Normally vertical surfaces are selected as locating datum and cylindrical faces are selected for clamping. Tolerance relations of the candidate datum feature with the machined surfaces in a setup are given importance. If no tolerance relationship exists between the surfaces, the surface with the largest diameter or the longest cylindrical surface having an orientation different from the surfaces being machined is selected as datum. Generally the two faces perpendicular to the axis of the part are selected as locating datum. In Fig. 2.2, for machining the features 5, 6 and 7 which have TAD right, the vertical face of feature 4 (which has the largest diameter) is selected as locating datum and the cylindrical face of feature 4 is used for clamping. The priorities used for selection of primary datum are as follows:

Priority 1: The face having the maximum number of tolerance relations with other features should be selected as primary datum. Huang and Liu [10] suggested several setup methods for attaining critical tolerance relationship between two features of a part. One of them is to use one feature as datum for machining the other

Fig. 2.2 Datum for a rotational part

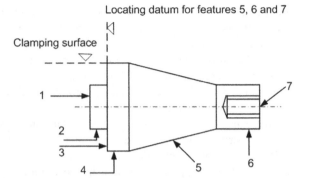

Locating datum for features 5, 6 and 7

Clamping surface

feature for attaining better tolerance relationship. For example, in Fig. 2.1, face 2 has the maximum number of tolerance relations with other features. It has parallelism tolerance with feature 1, perpendicularity tolerance with feature 8, and positional tolerances with features 4, 5, and 6. Therefore, face 2 is selected as primary datum for machining the features 1, 3, 4, 6, 7, and 8 in one setup.

Priority 2: Another priority for selecting primary datum is surface area of a face. The largest surface area face is normally selected as primary datum as it provides better stability during machining. However the selection is affected by orientation of the face, TAD of the features in the setup, etc. All the candidate faces for primary datum can be evaluated for surface area and the maximum area face can be selected.

Priority 3: Machined faces are selected as primary datum. The surface quality of datum is an important factor as it locates a component to be machined. Therefore, surface finish is one of the criteria for assessing the suitability of a face to be selected as datum.

For selecting secondary datum, all the faces perpendicular to the primary datum are considered and the largest face is selected as the secondary datum. Similarly, the tertiary datum is the largest face which is perpendicular to both primary and secondary datum.

2.5 Machining Operation Sequencing Within a Setup

In each setup, a number of features to be machined are grouped together. The appropriate machining operations to produce each feature are to be selected and sequenced in a proper and feasible manner. For example, drilling operation can be selected to produce a hole feature, milling operation can be selected to produce a step feature and so on. It may be necessary to consult the appropriate vendor catalogues of the manufacturing equipment present in the shop floor and manufacturing process handbooks for detailed information about process capabilities of various machining operations. These catalogues and handbooks provide the dimensions, tolerances and the surface finish ranges attainable by different machining processes. Sequencing these machining operations within each setup is the most challenging task in setup planning. Machining operations sequencing has the greatest impact on machined part accuracies. The decision making in sequencing machining operations depends on certain constraints, viz. precedence constraints, different machining constraints and good manufacturing practice. For example, machining of external surfaces is followed by machining of internal surfaces and rough machining is followed by semi-finish machining and then finish machining and so on. Similarly, boring (or reaming) must be performed after drilling, drilling must be performed before tapping threads in a hole. Grinding is usually the final operation to be performed in order to obtain the precision required of the feature. For external features, turning, taper turning and grooving are normally performed before grinding and so on.

One important criterion for machining operation sequencing is to minimize tool changes. By grouping the similar machining operations together, (for example, grouping all the drilling operations together) it is possible to reduce the number of tool changes and idle tool motion. The necessary knowledge for sequencing machining operations is based on heuristic and expert knowledge from various sources such as handbooks, textbooks and interviews with experts and skilled machinists. Some knowledge is gathered from observations of actual machining in the shop floor. Researchers have tried to generate feasible machining sequences using different approaches such as expert systems, fuzzy logic, neural networks, PSO techniques, etc. based on criteria of minimum number of setups and tool changes and non-violation of feature precedence relations [3–7, 12, 16].

2.5.1 Generation of Machining Precedence Constraints

During machining of the features comprising a part, certain precedence relations among the features are to be respected. These precedence relations arise due to basic manufacturing principles and feature interactions. A precedence relation between two features F1 and F2, denoted as F1 → F2, implies that F2 cannot be machined until the machining of F1 is complete. Different precedence relations are obtained due to area/volume feature interactions, tolerance relations, feature accessibility, tool interaction, fixturing interaction, datum/reference/locating requirements, and constraint of good manufacturing practice. Some examples of precedence constraints are as follows: if there is a feature a of name hole which is to be drilled on a chamfered face b, then due to tool interaction constraint, the drilling of hole a is to be done prior to the chamfer b, or if there is an internal feature a which is nested in another feature b, then due to parent-child precedence constraint, the machining of feature b is to be done prior to the machining of a. Similarly, if a feature a is the datum/reference for feature b, then a has to machined prior to b which will result in datum/reference precedence constraint. Figure 2.3 shows some of the precedence relations collected from the literature.

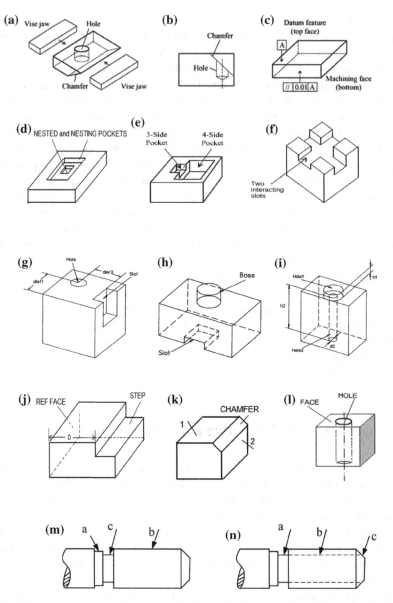

Fig. 2.3 Different precedence relations collected from the literature. Reproduced with kind permissions: **a–e** from Liu and Wang [16], Copyright [2007] Elsevier, part of **f** from Pal et al. [19], Copyright [2005] Elsevier and **g–i** from Zhang et al. [20], Copyright [1995] Springer Science and Business Media. **a** Drill hole → Chamfer. **b** Drill hole → Chamfer. **c** Datum A → Bottom face. **d** Nesting pocket → Nested pocket. **e** Base 4-side pocket → 3-side pocket. **f** Slot1 → Slot2 or Slot2 → Slot1. **g** Slot → Drill hole. **h** Slot → Boss. **i** Hole1 → Hole2. **j** Ref face → Step. **k** Faces 1 and 2 → Chamfer. **l** Face → Drill hole. **m** a → c, b → c. **n** a → b, c → b

Figure 2.3a depicts a precedence constraint arising due to fixturing interaction. Drilling the hole should precede the chamfer as fixturing will be difficult for drilling after chamfering. There will be less contact area for clamping the vise jaw if chamfering is done first. For similar reason, the slot precedes the boss in Fig. 2.3h. An accessibility/tool interaction constraint is shown in Fig. 2.3b where positioning the drilling tool will be difficult if chamfering is done first. Same is the case in Fig. 2.3m, where machining of the groove c between two adjacent external cylindrical surfaces a and b is done after machining of a and b. Figure 2.3c depicts the precedence constraint arising due to tolerance relation with the datum feature. The bottom face has tolerance relation with the datum face A and face A is to be machined first. Figure 2.3d shows two nested pockets having volumetric interaction, i.e. common volume to be removed. The smaller pocket is nested in the bigger pocket and the machining of the bigger/nesting pocket precedes the smaller/nested pocket. This type of precedence relation is called parent–child relation. The parent/nesting feature is to be machined prior to the child/nested feature. In Fig. 2.3e, the two pockets have only area interaction in the form of a common face. The 4-side base pocket is opening up to another 3-side pocket and the convention is to machine the base feature first. Figure 2.3f is a case of no precedence; any of the two slots can be machined first. Figure 2.3g, j shows the precedence of machining the reference features first. In (g), the hole is referenced with respect to the slot and in (j), the step is referenced with respect to the vertical face and reference features are to be machined first. Figure 2.3i shows good manufacturing practice of drilling the smaller depth hole prior to higher depth hole. Figure 2.3k, l shows the precedence of machining the adjacent faces first and then chamfering/drilling. There are certain constraints requiring that the subsequent features should not destroy the properties of features machined previously. An example is that the machining of a chamfer and a groove must be completed prior to that of the adjacent thread as shown in Fig. 2.3n.

These feature precedence relations are derived from manufacturing practice and there may be uncertainty about the validity of some assumed relations. The optimal machining sequence depends to a large extent on precedence relations. The validity of the precedence relations are to be reviewed keeping in mind the other related factors such as machining cost and time, work material properties, the required surface finish, machining passes (single or multi), etc.

First, a sequence of machining operations is created within a setup based on their precedence relations. This operation sequence can be modified by grouping operations of same tool together as long as the precedence relations are respected. Moreover, for machining operation sequencing within a setup, the information on preceding operation for each machining operation is required. For example, the preceding operation for machining a nested feature is machining of the nesting feature which is again preceded by machining of its reference feature. These information/facts are created by the generation of precedence relations. An operation may have multiple preceding operations. A machining operation is assigned to a setup only if all its preceding operations have been assigned. Thus, using the precedence constraint information, a feasible sequence of machining operations within each setup is generated. The machining operations are arranged in the sequential order in which they are to be performed.

2.5.2 Good Manufacturing Practice

Machining operations sequencing also depends on some rules of thumb evolving from decades of experience which are practised in the industry. These are considered as good manufacturing practice. For example, in case of drilling of two concentric holes, a hole of smaller diameter is drilled prior to a hole of larger diameter. Similarly, the hole of longer depth is drilled prior to the hole of shorter depth if they are concentric. However, some precedence relation may have an element of uncertainty. In the above mentioned examples of drilling concentric holes, the decision depends on many related factors like hole dimensions, ease of access, tool used, possibility of tool damage, material properties, cutting parameters, etc. Therefore, validity of the precedence relations are to be reviewed keeping in mind the other related factors.

2.6 Setup Sequencing

After the features and their machining operations within a setup are sequenced, the setups are also to be sequenced in a similar manner. Precedence relations described above are very important and prime criterion for setup sequencing. Moreover, for sequencing the setups, effect of machining of the features in the preceding setups on their successive setups are to be considered. A setup where greater numbers of features are present should not be considered first for machining. It may give rise to problems of instability and insufficient locating and clamping surface area for the remaining setups. For the same reasons, it is preferred that smaller sized features should be machined prior to larger sized features. Considering these constraints, the following principles can be followed for sequencing different setups for machining a component:

- Setups are sequenced depending on the precedence relations existing among the features present in different setups.
- The setup with the maximum number of features is preferably machined last provided precedence relations among the features are respected.
- Feature dimensions are to be taken into account and larger sized features are preferably machined last as they affect the stability, locating and clamping in subsequent setups.

2.7 Conclusion

In this chapter the different phases of setup planning are presented in detail. Feature grouping, setup formation, datum selection, machining operation sequencing and setup sequencing functions are discussed with relevant examples. Feature precedence relations arising due to various machining conditions are explained with examples. The role of feature precedence relations in machining operation sequencing and setup sequencing is highlighted.

References

1. Bansal S, Nagarajan S, Reddy NV (2008) An integrated fixture planning system for minimum tolerances. Int J Adv Manuf Technol 38:501–513
2. Deiab IM, Elbestawi MA (2005) Experimental determination of the friction coefficient on the workpiece-fixture contact surface in workholding applications. Int J Mach Tools Manuf 45:705–712
3. Gologlu C (2004) A constraint based operation sequencing for a knowledge based process planning. J Intell Manuf 15:463–470
4. Gu P, Zhang Y (1993) Operation sequencing in an automated process planning system. J Intell Manuf 4:219–232
5. Gu Z, Zhang YF, Nee AYC (1997) Identification of important features for machining operations sequence generation. Int J Prod Res 35:2285–2307
6. Guo YW, Li WD, Mileham AR, Owen GW (2009) Application of particle swarm optimization in integrated process planning and scheduling. Robot Comput Integr Manuf 25:280–288
7. Guo YW, Mileham AR, Owen GW, Maropoulos PG, Li WD (2009) Operation sequencing optimization for five-axis prismatic parts using a particle swarm optimization approach. Proc ImechE Part B J Eng Manuf 223:485–497
8. Hazarika M, Dixit US, Deb S (2010) Effect of datum surface roughness on parallelism and perpendicularity tolerances in milling of prismatic parts. Proc ImechE Part B J Eng Manuf 224:1377–1388
9. Hebbal SS, Mehta NK (2008) Setup planning for machining the features of prismatic parts. Int J Prod Res 46:3241–3257
10. Huang SH, Liu Q (2003) Rigorous application of tolerance analysis in setup planning. Int J Adv Manuf Technol 3:196–207
11. Huang SH, Xu N (2003) Automatic set-up planning for metal cutting: an integrated methodology. Int J Prod Res 41:4339–4356
12. Kim IH, Suh H (1998) Optimal operation grouping and sequencing technique for multistage machining systems. Int J Prod Res 36:2061–2081
13. Kim IH, Oh JS, Cho KK (1996) Computer aided setup planning for machining processes. Comput Ind Eng 31:613–617
14. Kumar AS, Nee AYC, Prombanpong S (1992) Expert fixture-design system for an automated manufacturing environment. Comput Aided Des 24:316–326
15. Lee NKS, Chen JY, Joneja A (2001) Effects of surface roughness on multi-station mechanical alignment processes. ASME J Manuf Sci Eng 123:433–444
16. Liu Z, Wang L (2007) Sequencing of interacting prismatic machining features for process planning. Comput Ind 58:295–303
17. Mei J, Zhang HC, Oldham WJB (1995) A neural networks approach for datum selection. Comput Ind 27:53–64
18. Ong SK, Nee AYC (1998) A systematic approach for analyzing the fixturability of parts for machining. ASME J Manuf Sci and Eng 120:401–408
19. Pal P, Tigga AM, Kumar A (2005) A strategy for machining interacting features using spatial reasoning. Int J Mach Tools Manuf 45:269–278
20. Zhang YF, Nee AYC, Ong SK (1995) A hybrid approach for setup planning. Int J Adv Manuf Technol 10:183–190

Chapter 3
Methods for Solving Setup Planning Problems

Abstract Automatic setup planning has been an active area of research for a long time. Traditional approaches of setup planning have been using decision tree, decision table, group technology, algorithms and graphs. A number of experts systems also have been developed for setup planning. Expert systems follow forward chaining or backward chaining. The forward chaining supports data-driven reasoning; whilst the backward chaining supports goal-driven strategy. A number of soft computing tools have also been used in setup planning. Prominent among them are fuzzy set, neural networks and evolutionary optimization. Fuzzy set theory takes care of uncertainty and imprecision in the information. Artificial neural network can learn from the experience. Evolutionary optimization techniques help to provide optimum or near optimum solution. The goal may minimize the number of setups, manufacturing cost or resource consumption. The popular evolutionary optimization methods are genetic algorithms, particle swarm optimization and ant colony optimization. Nowadays cloud computing is also finding its application in setup planning.

Keywords Methods of setup planning · Decision tree · Group technology · Algorithms · Graph theory, fuzzy set theory · Artificial neural network · Evolutionary optimization · Genetic algorithms · Particle swarm optimization · Ant colony optimization · Cloud computing

3.1 Introduction

Setup planning is the part of computer-aided process planning (CAPP) concerned with various phases such as feature grouping, setup formation, datum selection, machining operation sequencing, and setup sequencing. Solving as well as automating setup planning problem is regarded as one of the most important activities in CAPP. Although the efforts to automate setup planning have been going on since 1980s, it is still a complex task. A feasible, optimal and automatic setup

© The Author(s) 2015   41
M. Hazarika and U.S. Dixit, *Setup Planning for Machining*,
SpringerBriefs in Manufacturing and Surface Engineering,
DOI 10.1007/978-3-319-13320-1_3

plan involves a number of constraints that are mutually conflicting considering technological, geometrical and economical aspects of both design and manufacturing domain.

Automatic setup planning has been an active research area in the last three decades and it is widely investigated by various researchers. Automatic setup planning approaches found in the literature are of diverse nature in terms of objectives, constraints and techniques used. Many researchers have addressed this problem from different perspectives and proposed different methods of setup planning based on analysis of part geometry, tool approach direction, tolerance requirements, precedence constraint analysis, fixtures needed, and manufacturing resources. Future trends of setup planning and process planning, its industrial perspectives and models, and integration with product design and manufacturing are widely analysed. Various researchers have focussed on the ultimate goal, constraints and various sub-tasks of setup planning in search of optimal setup plans. Some important review articles on setup planning and process planning are found in [17, 19, 42, 58, 67, 69].

Traditional approaches such as decision trees, decision tables, group technology (GT), algorithms and graph theory have been used for solving the problem of setup planning. In the following sections, a brief discussion on the application of these approaches in process planning and setup planning is presented. Pros and cons of different approaches are also discussed.

3.1.1 Decision Tree, Decision Table and Group Technology (GT) Based Approaches

Decision trees and decision tables are useful decision making tools. A decision tree is a way to represent information and knowledge. Conditions (IF) are set as branches of the tree and predetermined actions (THEN) can be found at the junction/node of each branch. The condition specified on each branch must be satisfied in order to traverse that branch. If the condition specified on a branch is true, then that branch can be traversed to reach the next node and this process is continued until a terminal point on the tree is reached. If the condition specified on a branch is false, then another branch may be followed until the terminal point is reached. Figure 3.1 shows the structure of a decision tree used to decide weather to machine a face first and then drill hole (on that face) or vice versa. Decision is to be taken based on uncertain knowledge. When holes are drilled on a face, burrs are formed on the edge of the hole which affects the desired surface finish of the face. Therefore, decision is to be taken on condition, e.g. if drilling burr size is big, then drilling is succeeded by facing so that good surface finish is obtained, and if burrs are of negligible size facing may be done prior to drilling. If facing precedes drilling, surface finish of the face is to be checked. In case the surface finish is not satisfactory, finishing operation is performed on the face as shown in Fig. 3.1.

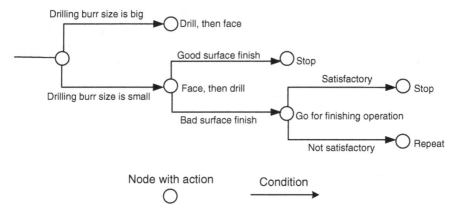

Fig. 3.1 Structure of a decision tree

The algorithm for implementing a decision tree may be written in any of the procedural programming languages such as FORTRAN, PASCAL, C, etc. Group Technology (GT) codes and special descriptive languages are used for representing the part description. Special descriptive language is used for part description in generative CAPP system AUTAP-NC [18]. GT uses similarities between parts to classify them into part families. In the context of machining process planning, a part family consists of a set of parts that have similar machining requirements. An appropriate classification and coding system is to be used for the entire range of parts produced in a shop. All the existing parts are coded following the adopted scheme for coding. Each part family is then represented by a family matrix. The next step is to prepare a standard process plan that can be used by the entire part family. The standard process plans are then stored in a database and indexed by family matrices. Example of some commercially available GT coding systems are Opitz coding, KK-3 coding, MICLASS coding, DCLASS coding, etc. An example of the application of decision tree in generative CAPP is Expert Computer-Aided Process-Planning (EXCAP) [10]. EXCAP generates process plans for machining symmetric rotational components. M-GEPPS is another CAPP system which uses decision tree for process planning of rotational parts [65]. KK-3 GT coding is used for providing part information as input to the system. Decision trees and decision support systems have long been used for process planning of both rotational and prismatic parts machined by operations such as turning, drilling, reaming, boring, slotting, milling, thread cutting, etc. Normally they are designed to perform coding and classification of parts, generate a list of required machining operations, select machine tools and cutting tools, optimize cutting parameters and provide alternative solutions.

Interfacing CAD system with a GT coding system is a common practice for generating feature information for process planning. There are various efforts to integrate CAD and CAPP with automatic GT coding leading to GT-based automated process planning. These systems normally allow a CAD system to be interfaced to a CAPP system through GT coding. Common neutral file data exchange standards like Standard for the Exchange of Product Model Data (STEP), Initial Graphics Exchange Standards (IGES) are used to connect dissimilar CAD packages to CAPP through a GT coding scheme. The details of the features are included in the codes. The final step is the generation of process plans for machining of the component. This is achieved through a program which is able to interpret the GT codes and generate an optimized process plan for machining of the component. Thus product designs with dissimilar formats from various CAD systems can be interconnected and automatically coded for multiple manufacturing purposes. Sometimes, developed coding scheme is integrated with knowledge-based systems (e.g. expert system) for the selection and sequencing of machining operations and selection of cutting tools. By integrating the proposed GT code and process planning knowledge, the machining facilities can be optimized. Readers are directed to the literature [30, 32, 39, 40, 49] for more insight into such systems.

Decision table is another tool used to represent process planning information. It organises the conditions (IF), actions (THEN) and decision rules in a tabular form. Conditions and actions are placed in rows of the decision table, while decision rules are placed in the columns. When all the conditions in a decision table are met, a decision is taken. The algorithm for implementing the decision table may be written in either some specially developed language or any of the procedural programming languages such as FORTRAN, PASCAL, C, etc. Decision tables are generally used in combination with algorithms or knowledge-based systems for decision making. The knowledge-base comprises machining rules and machine tool and cutting tool data. Rules are applied to determine the machining sequence, selection of cutting tools, cutting conditions for machining operations, etc. Table 3.1 is a sample decision table. Hierarchical and Intelligent Manufacturing Automated Process Planner (HI-MAPP) [4], and Cutting Technology (CUTTECH) [3] are two such automated process planning systems which use a combination of decision tables and knowledge-based rules for decision making.

Decision trees, decision tables, and GT codes, often used for traditional setup planning and process planning systems, work effectively only for simple decision making processes. The main limitation with the decision trees and decision tables is that they are relatively static in terms of representing the process planning knowledge. These are primarily methods to represent knowledge and are coded line by line in the program. Any modification to the current knowledge would require rewriting

Table 3.1 A sample decision table

Conditions	Rules			
Condition-1	Yes	No	Yes	Yes
Condition-2	No	No	Yes	Yes
Condition-3	Yes	Yes	No	No
Actions				
Action-1		X	X	
Action-2	X	X		X

of the original program. They lack the ability to automatically acquire knowledge and need longer response time. Moreover, GT code based input is not suited for automated process planning systems, since coding is a manual process. It is both time consuming and prone to error. Another major disadvantage of GT coding based systems is the cost involved in creating and maintaining databases for the part families.

3.1.2 Algorithmic and Graph Theoretic Approaches

Algorithms and graphs are powerful mathematical tools that have been used for solving setup planning problems. An algorithm is a sequence of finite logical and mathematical expressions for solving a given problem. A graph is a collection of finite number of vertices and edges. Each edge is identified with a pair of vertices. A setup planning problem can be formulated in terms of graphs. For example, two faces of a part may be represented by two vertices and the tolerance relation between the two faces can be represented by the edge connecting the two vertices. Again, machining operations can be represented as vertices and edges between two vertices can represent the precedence relations among those machining operations. An optimal setup plan can be searched by sequencing the machining operations so that none of the precedence constraints are violated. For traversing the required vertices or edges, different algorithms are used and the shortest path to the terminal point is searched. Common algorithms used for finding the shortest path through a graph are depth-first-search and breadth-first-search methods, gradient projection method, branch and bound method, etc. Generally, a searching strategy is developed which is used to search through the graph for feasible machining operation sequences for different setups to machine a part. These sequences are further optimised with respect to minimum number of setups, machine change and tool change, minimum machining time and machining cost. The criteria considered for setup planning are TAD, tolerance relations, precedence relations and orientation of the features. Both algorithmic and graph theoretic approaches have been used in CAPP for setup planning for the last three decades. Some important examples can be found in [22, 41, 56, 57, 72].

Attaining the specified tolerances is a crucial factor for the quality as well as functionality of a finished part. Tolerance is considered as the main driving factor in setup planning by many researchers. In these approaches, the precision of the final part is treated as the main criterion for setup planning. One important objective in setup planning is to ensure that the finished component meets design tolerance requirements. Therefore, design tolerances are used as the major guideline for setup planning for a feasible setup plan. The tolerance information of a part should be detailed so that process planning is proper in order to attain those tolerances. It is the task of process planning to select appropriate processes and machines to ensure that design tolerance requirements are met. A tolerance chart analysis can be done to verify if a process plan is capable to impart the design tolerance specifications to a part. In most of the tolerance based setup planning, the problem of identifying optimal setup plan is transformed into a graph search problem. Part information is represented by a feature and tolerance relationship graph (FTG) and setup information by datum and machining feature relationship graph (DMG). The problem of optimal setup planning is formulated as conversion of FTG to DMG based on tolerance analysis. FTG is constructed by representing each feature as a vertex and the tolerance relation between a pair of features by the edge connecting the vertices. Similarly, DMG is constructed with datum and feature information. Tolerance relations are used for machining operation and setup sequencing, datum selection, etc. Both dimensional and geometric tolerances are considered in these approaches which are used as critical constraints. The main principle followed is that tight tolerance features are to be machined in the same setup so that errors are minimized. A good setup plan should achieve the highest quality in terms of tolerances and be of the lowest cost. Various efforts of setup planning where tolerance is the key concern can be found in the literature. One important contribution is by Huang and Liu [26, 27] where the authors focussed on an important issue of tolerance analysis in setup planning. They developed a tolerance normalization method to express different tolerances in a common unit for comparing them. Normalized tolerance is an angle representing the maximum permissible rotation error when locating a component. Smaller the normalized tolerance, tighter is the tolerance between the features. Normalized values of different geometric tolerances among the features, e.g. parallelism, perpendicularity, angularity, position, concentricity, symmetry, etc. are considered for comparison. Features and their tolerance relations are represented with a tolerance graph. Readers are directed to [25, 28, 63, 66, 68, 71] for some examples.

Many researchers consider fixturing requirements as an integral part of setup planning. To provide a robust and practical solution for setup planning, setup planning has to be integrated with fixture planning by considering setup and fixturing requirements simultaneously. The pioneering work in setup planning integrated with fixturing was started by Boerma and Kals [6, 7]. A fixture provides some locating and clamping mechanism to support and maintain the work piece in a particular position in a setup and resist gravity and other operational forces. The importance of integrating setup planning and fixturing for precise machining cannot be overlooked. The purpose of setup and fixturing is to ensure the stability and precision of the workpiece during machining processes. Output of this approach is a combination of setup plans and fixturing solution for each setup. The main fixturing constraints considered in these approaches are machining forces, stability and restraint of the machined part, interference checking for fixturing, tool interference, locating and clamping faces and points, and minimum part deformation. These fixturing constraints along with TAD, precedence relations and tolerance can be considered to generate practical and feasible setup plans. However, most of these works in the literature deal with the conceptual fixture design phase by identifying the datum features. Fixture layout, machining force, clamping force and process parameters play vital role in formulating a setup plan considering the feasibility of fixturing. These aspects are to be given more importance for a better solution. Clamping force and machining force are two major factors which affect stability and deformation of workpiece and fixtures. In most of the works considering forces, only static force is considered. However, dynamic machining force which changes with time is the most affecting factor in practical machining works. The fixture driven approach when integrated with setup planning, can give practical solutions.

Adaptability of the setup plans to changing manufacturing environment is an important issue. Setup plans developed prior to actual production may become infeasible during actual production time due to changes in the conditions on the shop floor. Adaptation to the changing scenario is a crucial factor in this era of lean and agile manufacturing. It is observed that importance is shifted from stand-alone setup planning system to dynamic setup planning system in geographically distributed manufacturing environment. In the emerging trend of virtual manufacturing, a part is designed and manufactured in different sites using the facilities available in a multi-enterprise scenario. Java and Web technologies coupled with eXtensible Modeling Language (XML) file format provide means for the transfer of information between various manufacturing systems. The importance of dynamic and adaptable manufacturing software systems is manifold. Traditional software systems for automating setup planning are static in nature and they do not respond to the changes in the situation. Adaptive and dynamic setup planning and process planning systems are to be developed to cater to the need in the present manufacturing scenario. Some adaptable process planning approaches can be looked up in [2, 8, 20, 21, 31, 33, 61].

Approaches based on algorithms and graphs have been reported to give good and accurate results. However, there are limitations of these approaches. They are inflexible particularly for new situations. For example, if there is a change in the manufacturing environment, any modification of the current methodology would require rewriting of the original program. Moreover, for a complex problem, the size of the program becomes large and may need large computing resources. Because of these drawbacks, use of Artificial Intelligence (AI) techniques is gradually increasing in setup planning giving better results compared to traditional methods of setup planning.

3.2 Application of Artificial Intelligence and Soft Computing to Setup Planning

The traditional methods such as algorithms and graphs, decision trees, and decision tables suffer from various shortcomings. They are inflexible and lack the necessary intelligence to automatically acquire knowledge. To overcome some of these limitations, approaches based on artificial intelligence (AI) and soft computing techniques are explored by researchers. In the recent years, many researchers have incorporated AI and soft computing in setup planning and process planning. AI is defined as the simulation of human intelligence on a machine, so as to make the machine efficient to identify and use the right piece of knowledge at a given step of solving a problem [37]. AI problems are those problems which do not yield results from conventional mathematical or logical algorithms and can only be solved by intuitive approach. Expert systems, natural language processing, image recognition, and robotics are some of the application areas of AI. According to Prof. Zadeh, "soft computing is an emerging approach to computing, which parallels the remarkable ability of human mind to reason and learn in an environment of uncertainty and imprecision" [29]. Soft computing refers to a collection of tools and techniques which can model and analyze complex problems. Earlier computational approaches could model and precisely analyze only relatively simple problems. More complex problems arising in psychology, philosophy, medicine, computer science, engineering and similar fields often remained intractable to conventional mathematical and analytical methods. These problems are better solved by soft computing techniques. Unlike conventional methodologies (hard computing techniques), soft computing techniques are tolerant of imprecision, uncertainty and partial truth. They do not suffer from inflexibility of conventional algorithmic approaches. Soft computing covers a number of techniques such as artificial neural networks (ANN), fuzzy logic, evolutionary algorithms such as genetic algorithm (GA), simulated annealing (SA), ant colony optimization (ACO), particle swarm optimization (PSO), etc. These methods have the potentials to deal with highly non-linear, multi dimensional and complex engineering problems. The application of soft computing techniques is increasing with successful applications in different areas like engineering design, optimization, manufacturing system, process control, simulation and communication systems, etc. In the following sections, a brief discussion of the application of AI and soft computing techniques in setup planning is presented.

3.2.1 Expert System Based Setup Planning

An expert system is an AI tool used to solve problems that normally require human intelligence. Implementation strategies for problem solving methods should be general enough to capture knowledge from different sources and simple enough to provide an easily maintainable environment. Expert systems have been the most commonly adopted implementation strategy among the CAPP developers for their ability of reasoning, collection and representation of large amount of knowledge, and explicit inference route. It is a computer program that contains subject specific knowledge of one or more experts. Expert's knowledge in a specific field is collected and stored in the knowledge-base of the expert system. One of the important needs for expert system development is to capture human expertise and use it. Human expertise is scarce and is lost due to retirement, transfer, etc. An expert system analyzes the user supplied information about a specific problem and utilizes reasoning capabilities to draw conclusions. It emulates the problem solving and decision making capacities of a human expert. The different components of an expert system are:

- Knowledge-base containing domain specific knowledge collected from experts, books and manuals.
- Database containing declarative knowledge (facts) about the problem.

Fig. 3.2 Components of an expert system

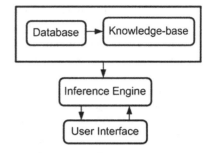

- Inference engine which is the reasoning mechanism that provides conclusions.
- User interface by which the user communicates with other components of the expert system.

Figure 3.2 shows the basic components of an expert system.

The modular nature of expert systems makes them easier to encapsulate knowledge and expand them by incremental development. Expert systems should be flexible because facts and rules in an expert system require constant updating. As new technology, equipment, and processes become available, the most effective way to manufacture a particular part also changes. An expert system stores knowledge in a special manner so that it is possible to add, delete, and modify knowledge in the knowledge-base without recoding the program. They can be adapted to some extent to the changing manufacturing environments by modifying the existing rules in the knowledge-base or by introduction of new rules. Separation of control knowledge or inference engine from the knowledge-base gives added flexibility to the expert systems. In general, expert system is ideally suited when the problem cannot be well defined analytically, the number of alternative solutions is large, the domain knowledge is vast, and relevant knowledge needs to be used selectively.

Knowledge-base in an expert system contains the domain specific knowledge that is used to solve problems. In case of setup planning, Knowledge-base contains manufacturing knowledge, knowledge about the manufacturing resources, machining processes, working material, machinability, etc. The collected knowledge can be represented and stored in the knowledge-base in various forms. One of the most commonly used forms of knowledge representation is IF–THEN rules. Setup planning knowledge is incorporated in the knowledge-base as production rules in the form of IF–THEN rules. An IF–THEN rule can be represented in the form: A → B. This is interpreted as 'IF condition A is satisfied THEN action B occurs'. The A portion is called the antecedent part and the B portion is called the consequent part of a rule. The knowledge is elicited by the knowledge engineer/system developer from various experts, books and manuals. Other methods of representing knowledge are semantic networks, conceptual graphs, frames, object oriented schemes and Petri nets. The rules are represented in natural language (e.g. English). Therefore the knowledge-base can be developed rapidly without the need to perform extensive programming. Adding, modifying, or deleting rules does not require extensive system changes. Feature tolerance relationship, TAD, precedence constraints, datum and reference requirements are some of the constraints considered for setup planning.

The database contains a set of facts about the part to be manufactured, e.g. part geometry, machining operations, dimensions and tolerances of the part, and machine tool and cutting tool information. These data files (facts) are given by the user as input to the expert system. The facts are used to match against the IF parts of the rules stored in the knowledge-base. The information needs to be converted from its original data format to the representation format for facts supported by the expert system. The database also contains mathematical functions and external programs that are necessary for performing different calculations. The inference engine contains and realizes the decision making strategy. It links the rules in the knowledge-base and the facts in the database to form a line of reasoning for drawing inference and generate setup plans. It is an independent module that makes the expert system more flexible. There are two methods of reasoning by which the inference engine arrives at some decision, forward chaining and backward chaining. Forward chaining supports data-driven reasoning. In this method, the inference engine starts from a set of conditions and moves towards a conclusion. It tries to match the available facts from the database with the antecedent part of the rules in the knowledge-base. When matching rules are found, the consequent parts of the rules are executed. Thus new facts are generated which in turn cause other rules to fire. This process continues until no more matching rules are available. For example, for machining a component, forward chaining strategy starts with the initial blank and matches the facts and data about the features of the part and the machining operations with the production rules in the knowledge-base until the final finished part is reached. An example of setup planning for prismatic part using forward chaining with CLIPS expert system shell is presented in Chap. 4. GARI [14], AGFPO, Semi Intelligent Process Planning (SIPPS), DOPS and FEXCAPP are some process planning systems that use forward chaining reasoning strategy [36]. Backward chaining supports goal-driven reasoning. The conclusion/goal is known and the inference engine checks the consequent parts of the rules in the knowledge-base to find a matching for the goal. If matching is found, facts are searched to cause any of those rules to fire. In context of machining a component, backward chaining reasoning starts with the finished component and matches the consequent part of the rules to the facts and data about the features of the part and the machining operations until the terminal point (in this case, the initial blank) is reached. Technostructure of Machining (TOM), Cutting Technology (CUTTECH), EXCAP process planning systems use backward chaining reasoning strategy [36]. The user interface is the means of communication between a user and the expert system. Input to the expert system, output of the final results and decisions are communicated through this interface.

The methods for developing expert systems can be classified into two types, viz., programming from scratch and using expert system shells. The use of programming provides more flexibility and control to the developer. Artificial intelligence languages such as List Processing (LISP), Programming in Logic (PROLOG) and conventional programming languages such as C, Pascal, Java, etc. can be used for developing expert systems. However, developing a complete expert system using programming requires greater expertise and tremendous amount of time and work. To simplify the task of expert system development, expert system shells are used. An expert system shell is a software system where the developer has to build the knowledge-base. It contains a built-in inference engine, a user interface, a set of knowledge representation structures and facilities to interface with external systems. Some examples of expert system shells are Empty MYCIN (EMMYCIN), Expert System Shell (EXSYS), C Language Integrated Production System (CLIPS), Automated Reasoning Tool (ART), G2, LEVEL5, etc. [50]. Using expert system shells can speed up the expert systems development time. For some recent expert system based setup planning, the readers are directed to [12, 22, 35, 44, 45, 60].

The expert systems, however, suffer from some weaknesses. It is restricted to the fields where expert knowledge is available and it is unable to infer when information provided is incomplete. It cannot automatically acquire knowledge and lacks the ability to deal with uncertainty. The new knowledge must be incorporated into the expert system by specifying it in explicit rule format.

3.2.2 Fuzzy Logic Based Setup Planning

One important AI technique used for setup planning is fuzzy logic. Fuzzy logic is best used to deal with reasoning under uncertainty. It is able to handle uncertainty and reason with imprecise information. As decision making in setup planning involves the use of uncertain knowledge to a large extent, use of fuzzy logic can help to get a better solution. The application of fuzzy logic offers the advantage of structured knowledge representation (similar to that of expert systems) in the form of rules with linguistic labels. With fuzzy logic, the precise value of a variable is replaced by a linguistic description, which is represented by a fuzzy set, and inference is drawn based on this representation. The benefit of using fuzzy logic is the ability to solve practical, real world problems, which invariably involve some degree of imprecision and uncertainty. The procedural knowledge in fuzzy logic based systems is expressed as fuzzy IF–THEN rules. Either the antecedent part (IF part) or the consequent part (THEN part) or sometimes both the parts of a rule has fuzzy sets. Fuzzy logic and fuzzy sets have been used in CAPP for automating process planning, setup planning, selection of cutting parameters for machining, etc. A background of fuzzy sets is presented in the following sections.

3.2.2.1 A Background on Fuzzy Sets

Fuzzy set theory was first proposed by Zadeh [70]. Fuzzy set theory derives its motivation from approximate reasoning. With the introduction of fuzzy set theory, the scope of traditional mathematical approach is widened to accommodate partial truth or uncertainty. Transition from crisp (true/false) mathematics to fuzzy mathematics by means of fuzzy set theory has enabled computing with natural language. In fuzzy sets, the precise value of a variable is replaced by a linguistic variable. Linguistic variables can have linguistic values. If *temperature* is a linguistic variable, then its linguistic values can be *high*, *low* and *moderate*. These values are represented by fuzzy sets. Fuzzy sets can be used in human decision making process to draw conclusions from vague, ambiguous or imprecise information.

A set is a collection of elements. In a crisp set, the elements of the universe are either a member or non-member of a set. Fuzzy sets are those sets whose boundaries are vaguely defined. Fuzzy set theory may be considered as an extension of classical set theory. Classical set theory deals with crisp sets with sharply defined boundaries, whereas fuzzy set theory is concerned with fuzzy sets whose boundaries are imprecisely defined. The benefit of replacing the sharply defined boundaries with the imprecisely defined boundaries is the strength in solving real world problems, which involve some degree of imprecision and vagueness. An element of the universe may be a member of a fuzzy set to varying degrees. The same element can be a member of different fuzzy sets with different degree of membership. Unlike classical set theory, fuzzy set theory is flexible and focuses on the degree of being a member of a set. In a fuzzy set, the members are allowed to have any positive membership grade between 0 and 1. The membership grade is defined as the degree of being a member of a fuzzy set. Membership grades are subjective, but not arbitrary. For example, consider that there are two fuzzy sets 'young man' and 'old man'. A 25 years old person may be given a membership grade of 0.8 in the set of 'young man' by one expert and 0.9 in the same set by another expert. The same person may be given 0.2 and 0.3 membership grades in the set of 'old man' by the two experts respectively. Both these values of membership grades are considered reasonable. The slight difference is due to the difference in perception of the two experts. However, in a crisp set, the person will have membership grades 1 and 0 in the sets of 'young man' and 'old man' respectively. In a fuzzy set, a membership grade 1 indicates full membership and 0 indicates full non-membership in the set. Any other membership grade between 0 and 1 indicates partial membership of the element in the set. Some skill is needed to form a fuzzy set that properly represents the linguistic name assigned to the fuzzy set.

The process by which the elements from a universal set X are determined to be either members or non-members of a crisp set can be defined by a characteristic or discrimination function. For a given crisp set A, this function assigns a value $\mu_A(x)$ to every $x \in X$ such that

$$\mu_A(x) = \begin{cases} 1 & \text{if and only if } x \in A, \\ 0 & \text{if and only if } x \notin A. \end{cases} \tag{3.1}$$

Thus the function maps elements of the universal set X to the set containing 0 and 1. This is indicated by

$$\mu_A: X \to \{0, 1\}. \tag{3.2}$$

This kind of function can be generalized such that the values assigned to the elements of the universal set X fall within a specified range. These values are called membership grades of the elements in the set. Larger values denote higher degree of membership and vice versa. Such a function is called a membership function μ_A by which a fuzzy set A is usually defined. The fuzzy membership function is indicated by

$$\mu_A: X \to [0, 1] \tag{3.3}$$

where [0, 1] denotes all real numbers between 0 to 1 including 0 and 1.

To assign suitable values of membership grades and constructing the membership function is one of the most challenging tasks of fuzzy set theory. Design of fuzzy membership functions greatly affects a fuzzy set based inference system. Membership functions are subjective but not arbitrary. Normally an expert's opinion is sought to construct the membership function for a fuzzy variable. The geometrical shape of the membership function characterizes the uncertainty in the corresponding fuzzy variable. There are different types of membership function, e.g. triangular, trapezoidal, Gaussian function, S-function, π-function, etc. For ease of computation, linear membership functions such as triangular and trapezoidal functions are preferred. However, in order to mimic real life problem, non-linear membership functions may be used. A typical triangular membership function is shown in Fig. 3.3 [24]. It is constructed by taking the membership grade as 1.0 at most likely (m) and 0.5 at low (l) and high (h) estimates of a fuzzy parameter. The vertices l' and h' denote the extreme low and extreme high estimates of the parameter.

3.2.2.2 Fuzzy Set Operations

Fuzzy sets have been extensively used in decision making by employing various operations on fuzzy set theory. Fuzzy union ($A \cup B$) and fuzzy intersection ($A \cap B$) between two fuzzy sets A and B are the two most commonly used fuzzy

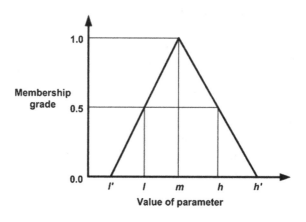

Fig. 3.3 A triangular membership function (with kind permission from Hazarika et al. [24], Copyright [2010], Springer Science and Business Media)

operations. The union of two fuzzy sets A and B, i.e. $A \cup B$ is defined as a set in which each element has a membership grade equal to the maximum of its membership grade in A and B. Similarly, the intersection of two fuzzy sets A and B, i.e. $A \cap B$ is defined as a set in which each element has a membership grade equal to the minimum of its membership grade in A and B. Fuzzy union and intersection operations are expressed as

$$\mu_{A \cup B} = \max\{\mu_A, \mu_B\},$$
$$\mu_{A \cap B} = \min\{\mu_A, \mu_B\}. \tag{3.4}$$

Fuzzy complement, fuzzy absolute difference, fuzzy product are some more examples of fuzzy set operations. Fuzzy set operations are useful in taking decisions in the presence of conflicting and incommensurable objectives [15]. For example, consider that a certain product requires functionality as well as aesthetic appeal as its attributes. Now, if the product has a membership grade of μ_1 in the set of 'functionality' and a membership grade of μ_2 in the set of 'aesthetic appeal', then the overall membership grade μ_o in the set of 'suitable product' can be found as

$$\mu_o = \min(\mu_1, \mu_2) \tag{3.5}$$

Here the overall performance of the product is dependent on the most poorly performing attribute. This type of strategy is called non-compensating strategy. An overall membership grade based on a compensating strategy may be defined as

$$\mu_o = \sqrt{\mu_1 \mu_2} \tag{3.6}$$

A weighted combination of the two strategies may also be considered. If the membership grades of a product in n different objectives are $\mu_1, \mu_2, \ldots, \mu_n$, then overall membership grade is defined as

$$\mu_o = (1 - \alpha)\min(\mu_1, \mu_2, \ldots, \mu_n) + \alpha \sqrt[n]{\mu_1 \mu_2 \ldots \mu_n}, \tag{3.7}$$

where α is a weight factor. Putting $\alpha = 1$ in Eq. (3.7), a pure compensating strategy is obtained and $\alpha = 0$ in Eq. (3.7) provides a pure non-compensating strategy.

3.2.2.3 Linguistic Variables and Hedges

Fuzzy set theory uses natural language and thus deals with linguistic variables. Linguistic variables can have linguistic values. If *age* is a linguistic variable, then its linguistic values can be *young*, *middle-aged* and *old*. A linguistic variable is often associated with fuzzy set quantifiers called hedges. The function of hedges is to modify the membership function of an already defined fuzzy variable. The examples of some hedges are *very*, *usually*, *fair*, *quite*, etc. Linguistic value of a fuzzy set can be modified by applying a hedge as an operator on the fuzzy set. For example, the hedge *very* performs concentration by reducing the membership values of the members and creates a new subset as shown below:

$$\mu_A^{very}(x) = [\mu_A(x)]^2 \tag{3.8}$$

Similarly, more or less, i.e. fair performs dilation and increases the degree of membership of fuzzy variables as follows:

$$\mu_A^{more\ or\ less}(x) = \sqrt{\mu_A(x)} \tag{3.9}$$

Thus, using linguistic hedges, computation can be done using natural language. Usually, a fuzzy set based model is built by using expert knowledge in the form of linguistic rules. Fuzzy set theory, compared to other mathematical theories is easily adaptable. Fuzzy sets allow possible deviations and inexactness in defining an element. Therefore fuzzy set representation suits well the uncertainties encountered in practical life.

From the above discussion on fuzzy sets, it is clear that use of fuzzy sets to deal with the uncertainty in setup planning is justified. Setup planning requires in-depth manufacturing knowledge needed for actual production. Decision making in setup planning problem involves the use of uncertain knowledge to a large extent. Manufacturing a part involves a number of decisions such as design interpretation, selection of material, selection of machining operations, selection of datum for a setup, machine tools and cutting tools, determination of cutting conditions and so on. Traditionally in manual planning, the process planner takes the above decisions, based on his/her intuition and rules of thumb gained from his/her experience. The knowledge for formulating the rules is based on heuristic and expert knowledge from various sources such as interviews with experts and skilled machinists, handbooks and textbooks. However, there may be uncertainty in the collected knowledge which affects the final outcome. Incomplete information, imprecision and vagueness in the acquired knowledge lead to uncertainty. Generation of a feasible and optimal setup plan depends to a large extent on the way uncertainty is managed by the system. Fuzzy set theory can be used to deal with such uncertainties. For example, precedence relation between machining of two features may depend on many uncertain factors. This fact was realized by Ong and Nee [51–55] and they applied fuzzy set theory to deal with the uncertainty associated with feature precedence relations. They used the concept of feature dependency grades to deal with uncertain feature relations. The dependency grade is basically a membership grade ranging from 0 to 1. If the dependency grade from feature A to feature B is 1, then feature A can be machined only after machining feature B. If the dependency grade is less than 1, then it is preferred that A should be machined before B, the strength of preference being proportional to dependency grade. There may be uncertainty on the shop floor, viz. resource and capacity constraints, machine breakdown, and tool failure that can be dealt with fuzzy sets. Another example may be the selection of datum for a setup. The decision on selecting a suitable datum for a setup depends on various factors like feature tolerance relationships, surface area of a face, its orientation, symmetry, and surface quality. As choosing the proper criteria for selecting datum is based on uncertain knowledge, fuzzy set theory can be used to deal with the uncertainty associated with datum selection for a setup. Fuzzy sets are used for machining operation selection, operation sequencing, setup sequencing under uncertainty.

Fuzzy decision making strategy based on the objectives of minimum number of setups, machining steps, machining time, manufacturing resources and cost has been used by various researchers for process planning and setup planning [20, 21, 62, 64]. A hybrid of fuzzy set theory with neural network or knowledge-based system is also used for setup planning and reported to give good results.

The main weaknesses of fuzzy set based methods are that they are restricted to the fields where expert knowledge is available and are unable to automatically acquire knowledge. For a fuzzy input or output variable, membership grades are assigned to map numeric data to linguistic fuzzy terms. Design of fuzzy membership functions/membership grades greatly affects a fuzzy set based inference system. The problem of finding appropriate membership functions/membership grades for the fuzzy variables poses a challenge to the researchers.

3.2.3 Artificial Neural Network (ANN) Based Setup Planning

Another AI technique used for setup planning is artificial neural network (ANN). ANN has been used by many researchers for setup generation, datum selection, feature sequencing and setup sequencing. ANN approaches offer number of advantages such as capability to automatically acquire knowledge from exemplars, higher processing speed, etc. Moreover, it is capable of adapting to changing environment through retraining. Because of the above mentioned advantages, ANN has been aptly used for setup planning and process planning. A few researchers tried an approach of mapping setup planning problem to the travelling salesman problem (TSP) using ANN. The feature sequencing in each setup is mapped to the well-known TSP problem by considering each feature as a city and the setup time (setup, fixturing and tool change time) as the distance between the cities. TSP is a combinatorial optimization problem studied in operational research and computer science. Given a list of cities and their pair wise distances, the task is to find a shortest possible tour that visits each city exactly once. Some recent examples of using ANN for setup planning can be found in [1, 9, 11, 47]. A background of artificial neural network is presented in the following sections.

3.2.3.1 A Background on Artificial Neural Network

An artificial neural network is an information-processing model that is inspired by the way human brain process information. It can be defined as a model for reasoning based on the human brain. A human brain can deal with a lot of complex information and perform parallel information-processing. In ANN, information is stored and processed in a similar manner. ANN having artificial intelligence behaves like a biological neural system. In biological neural system, the neurons receive electrical signals from other neurons, whereas in the artificial neural system, these electrical signals are represented as numerical values. In ANN, various nodes called neurons are interconnected in the network. These neurons work together to solve complex problems. They are used to find out the relationship between input/independent and output/dependent variables. Input signals are given to the neurons in the network and after processing, they provide output signals. ANN can learn the complex relationships inherent in the provided data and it is very useful in modelling complex processes for which mathematical modelling is difficult. Neural network has the ability to extract patterns from complex and imprecise data and detect trends that are not noticeable by human beings. It is robust compared to expert systems, fuzzy logic and other traditional methods. Even if the input data presented to ANN is incomplete or erroneous, it can still function by retrieving the relationship between the input and output data and generate the correct outputs. This is particularly useful in the problems where a number of input decision variables are involved. A trained neural network can be considered as an expert and used to provide predictions for unknown problems.

3.2.3.2 Topology of Artificial Neural Network

Topology means the different architectures of the artificial neural networks. To find out the relation between input and output variables in ANN, different types of network architectures are used. They can be categorized into the following types:

- Feedforward neural network
- Feedback neural network and
- Self organizing neural network

Of these, feedforward neural network is the most popular and most widely used. Multi-layer perceptron (MLP) neural network and radial basis function (RBF) neural network are two common types of feedforward neural network. Generally, an artificial neural network consists of an input layer, one (or more) hidden layer, and an output layer. Figure 3.4 shows a feedforward neural network architecture. Here, only one hidden layer is shown. In RBF neural network, only one hidden layer is present, but an MLP neural network can have more than one hidden layers.

Neurons in the input layer take the values corresponding to the different variables representing the input pattern. The input signals move from one layer to another layer in a forward direction. The output neurons of the preceding layer become input to the neurons in the succeeding layer. In a biological neural

Fig. 3.4 A feed-forward neural network architecture

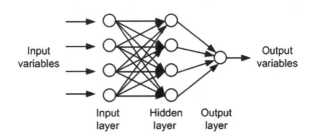

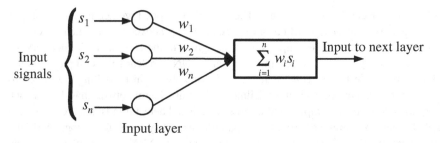

Fig. 3.5 Weighted sum of input signals in a neural network

network, electrical signals of varying intensity are sent to the neurons. In ANN, this phenomenon is modelled by multiplying each input by a weight value. Weighted sum of the inputs is calculated to find the total strength of the input signals. A neuron receives these input signals and provides an output signal depending on the processing function of the neuron. In Fig. 3.5, S_1, S_2, S_3, up to S_n are the n input signals received by an artificial neural network. The weights assigned to these signals are W_1, W_2, W_3, up to W_n. The weighted sum of these input signals is given by

$$S = \sum_{i=1}^{n} W_i S_i \tag{3.10}$$

This weighted sum is transferred to the next layer of neurons where the signal intensity is proportional to the weights. The second layer is called a hidden layer because its outputs are kept hidden internally. There may be more than one hidden layer present in a network. The final layer of the network is the output layer and the values of the neurons of the output layer constitute the result of the neural network. The total number of layers, number of neurons, and number of hidden layers needed are some of the important parameters for the design of the network architecture.

3.2.3.3 Training of Artificial Neural Network

The performance of an artificial neural network is judged from its ability to respond to input patterns and give correct output. To achieve this goal, the neural network is to be trained to respond correctly to a given input pattern. For training a neural network, the weights attached to the signals are to be adjusted so that the error between the predicted and the actual correct output is minimized. It is an iterative process that adjusts the weights of the neural network until the network can produce the desired output. The process of training a neural network is classified into the following categories:

- Supervised learning
- Unsupervised learning
- Reinforcement learning

Training of a feed forward neural network is a two step process. In the first step, the input signals propagate from input to output. As an input pattern is presented to the neural network, output is calculated layer by layer in the forward direction. Weighted sum of the signals are considered. The output signals of the preceding layer become input to the succeeding layer. This process is continued until the final output layer is reached. The final output is then compared with the desired output value and error for each output neuron is computed. In the second step, the computed error signals are fed backward through the network. These error signals are used to adjust the weights in the immediately preceding layer, and then the next preceding layer and so on till input layer is reached. Thus the error propagates backward and the procedure of weight adjustment is continued in such a way that the error between the desired output and the actual output is reduced. Training of a neural network following supervised learning is done in this manner. The algorithm used for adjusting the weights is called backpropagation algorithm which is most commonly used by the researchers. The backpropagation algorithm supplies the neural network with the input patterns and desired output/target patterns, which together constitute the training examples. The training of the neural network using supervised learning is applicable to problems where both input pattern and output/target patterns are known.

Unsupervised learning method of training is especially applicable where the target pattern of the problem is unknown. In the unsupervised learning process, the network is provided with a data set containing input patterns only. The unsupervised learning algorithm finds out hidden patterns among the data. Networks that are capable of inferring pattern relationships without getting any outside information are called self-organizing networks. Reinforcement learning process is an intermediate form of the above two methods of learning. Reinforcement learning is learning in which actions are taken for maximizing a numerical reward signal. The learner discovers which actions yield the most reward by trying various actions.

From the above discussion on artificial neural network, it is understood that it is well suited for setup planning problem for its capability to automatically acquire knowledge from examples, adapting to changing environment through retraining, and higher processing speed. For a setup planning problem of clustering features into different setups, the input pattern to the neural network may consists of all the features of the part along with their TAD, precedence relationships, tolerance relationships, and the set of tools needed for processing each feature. After running the neural network, the features are automatically clustered into a number of setups in a way to minimise the number of tool changes. For the problem of datum selection for a setup, the input to the neural network consists of the geometry of the part, and the tolerance specifications between the different part surfaces. The output of the network gives the surfaces selected for locating and clamping. However, the approaches based on neural networks suffer from some shortcomings too. The requirement of huge amount of training and testing data, the presence of the outliers and lack of one well-established standard method of training of neural network poses a challenge for the researchers. Moreover, neural networks provide no explanation of the rationale behind their inference procedure. Their lack of explicitly stated rules and vagueness in knowledge representation leads to a black box nature.

3.2.4 Application of Evolutionary Algorithms to Setup Planning

Setup planning is a discrete optimization problem involving simultaneous optimization of several objectives. Usually there is no single optimal solution, rather a set of alternative solutions exists. A combinatorial optimization problem cannot be solved by deterministic algorithms or the traditional optimization methods as they are either too time consuming or too difficult to find an acceptable solution. Evolutionary algorithms like genetic algorithm (GA), simulated annealing (SA), particle swarm optimization (PSO), ant colony optimization (ACO), etc. seem to be suited for this kind of problem because they process a set of feasible solutions in parallel, and search for multiple near optimal solutions. Evolutionary algorithms are intelligent and iterative search methods that mimic the process of natural biological evolution in specific steps and the social behaviour of species. Such species follow the specific steps of learning, adaptation, and evolution. Evolutionary algorithms share a common solution approach for solving problems. In the first phase, the problem is represented using a suitable format. In the second phase, the evolutionary search algorithm is applied iteratively to arrive at an optimum solution. Evolutionary algorithms are best suited to solve a particular class of problems called NP-hard problems. A problem is NP-hard if an algorithm for solving it can be translated into one for solving any NP-problem (nondeterministic polynomial time) problem. For NP-hard problems, no short method or algorithm exists that can give a simple and rapid solution. An optimal solution can be found only by testing all possible outcomes through exhaustive analysis. An example of a NP-hard problem is the well-known travelling salesman problem (TSP). TSP is a combinatorial optimization problem where the task is to find a shortest possible tour that visits each city exactly once, given a list of cities and their pair wise distances. Finding the optimal setup planning and process planning solutions are considered as NP-hard problems.

Genetic Algorithm (GA) is the first evolutionary-based technique developed and it is based on the Darwinian principle of the 'survival of the fittest'. GA follows the natural process of evolution through reproduction and ideally suited to parallel computing. GA can be effectively applied to combinatorial optimization problems where a small change leads to nonlinear behaviour in the solution space. GA can search very large solution space using probabilistic transition rules instead of deterministic ones. In GA, a solution is represented in the form of a string, called 'chromosome' which consists of a set of elements called 'genes', that represent the solution variables. The working cycle of GA is as follows: GA starts with a random population of solutions (chromosomes) and the fitness of each chromosome is evaluated against an objective function. To simulate the natural survival of the fittest process, best chromosomes undergo genetic evolution, i.e. they exchange information through crossover or mutation operations to produce offspring chromosomes. The offspring solutions are then evaluated and the best solution is selected. The process is continued for a large number of generations to obtain a near optimum solution.

GA has been mainly used for optimization of process plans and setup plans [5, 13, 43, 59]. The objective function is generally minimum number of setups/ minimum manufacturing cost/optimum resource consumption. Different constraints considered are machining time, range of cutting parameters, cutting power requirement, tolerance and surface finish, tool life, etc. There are many variants of GA. Nowadays, real coded GA is popular for solving optimization problems with real decision variables. However, in setup planning problems are of combinatorial nature and binary coded GA is the most suitable for them.

Despite its benefits, GA requires long processing time to generate an optimum solution. In search of better methods to reduce processing time and improve the quality of solutions, other evolutionary algorithms have been developed during the past decades. Particle swarm optimization (PSO) was developed by Kennedy and Eberhart [34] and PSO mimics the behaviour of a flock of migrating birds heading for an unknown destination. In PSO, each solution is a 'bird' in the flock and is called a 'particle'. Similar to PSO, another evolutionary method is ant colony optimization (ACO). ACO is a population based, general search technique which is inspired by the pheromone trail laying behaviour of real ant colonies for the solution of difficult combinatorial problems. ACO was developed by Dorigo et al. [16]. ACO mimics the behaviour of ants in trying to find the shortest route between their nest and a source of food. This is achieved from the pheromone trails that the ants deposit in the path they travel. The pheromone trail is considered as a means of communication to trace the path followed. Both PSO and ACO are tried recently for setup planning and process planning. Mohemmed et al. [48] explored the application of PSO to solve shortest path (SP) routing problems. SP problem can be related to setup planning in the context of machining operation sequencing in machining a component. The shortest path (SP) problem concerns with finding the shortest path from a specific origin to a specified destination in a given network while minimizing the total cost associated with the path. Guo et al. [23] investigated the applications of PSO in operation sequencing problem in setup planning. A comparative study among the modified PSO, GA and SA is presented highlighting their different characteristics. Lv and Zhu [46] defined process planning as a path searching problem and ACO is used to find the optimal path using the workshop resources with minimum time and cost. In the newly developed ACO, real distances are used to guide the ants in place of local pheromone deposits. Objective function is to search for a path to reach finish state from initial state with minimum time and cost. Krishna and Rao [38] used ACO to find the optimum operation sequence in a setup for machining of prismatic parts.

CAPP systems independent of specific operating system are also developed which can be used by geographically distributed manufacturers through Web and

Java technology. The Web based system provides a convenient platform for users to view and evaluate a design model effectively. Nowadays, cloud computing is finding application in manufacturing and it can be applied to CAPP also.

The difficulties associated with using mathematical optimization techniques to large scale engineering problems have contributed to the development of alternative solutions. To overcome these difficulties, researchers have proposed evolutionary-based algorithms for searching near-optimum solutions. Some researchers suggest that multi-objective search and optimization is an area where evolutionary algorithms do better than other optimization methods.

3.3 Conclusion

In this chapter the different approaches of solving setup planning problem are discussed in detail. From the above discussion, some critical observations are made as given below.

- It is observed that although there is success in automating some portions of process planning, there is not yet an automatic setup planning system fulfilling the requirements of commercial application. Absence of a standard definition of optimality of setup plans may be the reason behind this. A universally accepted definition of optimal setup plan, its evaluation criteria, a common scale for comparing optimality of different setup plans are to be formulated. For optimizing setup plans, the most commonly used objectives are the minimum number of setups/minimum setup time/minimum setup cost.
- For setup planning problems, use of AI techniques is gradually increasing, giving better results compared to traditional methods of setup planning. Some of the important soft computing techniques used by researchers are artificial neural networks, fuzzy logic, expert systems, evolutionary algorithms such as genetic algorithm, simulated annealing, ant colony optimization, particle swarm optimization, etc. Expert systems are good at logic, neural network is good for acquiring knowledge from examples, evolutionary algorithms are advantageous for optimization, and fuzzy logic is best used for dealing with reasoning under uncertainty.
- Various methods have been used for setup planning giving prime importance to tolerance achievement, fixturing requirements, optimization of setup plans, etc. Tolerance driven approaches ensures good quality of the part, but only tolerance consideration alone can not generate optimal setup plans unless other constraints are also considered. Fixture driven approaches, with proper integration of setup planning and fixturing can give practical solutions. Constrained optimization approach may be a promising venture for future research to find a truly global optimal setup plan considering all possible solution space.
- Adaptability of the setup plans to changing manufacturing environment is an important issue. Traditional software systems for automating setup planning are static in nature and they do not respond to the changes in the situation. There is

a need to develop an adaptive and dynamic setup planning system. Therefore, importance is shifted from stand-alone setup planning system to dynamic setup planning system in geographically distributed manufacturing environment. In the emerging trend of agile and virtual manufacturing, a part is designed and manufactured in different sites using the facilities available in a multi-enterprise scenario. Java and Web technologies provide means for the transfer of information between various manufacturing systems.

References

1. Amaitik SM, Kilic SE (2007) An intelligent process planning system for prismatic parts using STEP features. Int J Adv Manuf Technol 31:978–993
2. Azab A, ElMaraghy HA (2007) Sequential process planning: a hybrid optimal macro-level approach. J Manuf Syst 26:147–160
3. Barkocy BE, Zdeblick WJ (1984) A knowledge based system for machining operation planning. In: Proceeding of AUTOFACT, vol 6, Arlington, Texas, pp 2.11–2.25
4. Berenji RH, Khoshnevis B (1986) Use of artificial intelligence in automated process planning. J Comput Mech Eng 5:47–55
5. Bo ZW, Hua LZ, Yu ZG (2006) Optimization of process route by genetic algorithms. Robot Comput Integr Manuf 22:180–188
6. Boerma JR, Kals HJJ (1988) FIXES, a system for automatic selection of set-ups and design of fixtures. Ann CIRP 37:443–446
7. Boerma JR, Kals HJJ (1989) Fixture design with FIXES: the automatic selection of positioning, clamping and supporting features for prismatic parts. Ann CIRP 38:399–402
8. Cai N, Wang L, Feng HY (2008) Adaptive setup planning of prismatic parts for machine tools with varying configurations. Int J Prod Res 46:571–594
9. Chen CLP, LeClair SR (1994) Integration of design and manufacturing: solving setup generation and feature sequencing using an unsupervised-learning approach. Comput Aided Des 26:59–75
10. Davis BJ, Darbyshire IL (1984) The use of expert systems in process planning. Ann CIRP 33:303–306
11. Deb S, Ghosh K (2005) Application of artificial neural network and expert system for intelligent computer-aided process planning of rotationally symmetrical machined components. In: Proceedings of All India seminar on recent advances in manufacturing technologies, NIT Rourkela, India
12. Deb S, Ghosh K (2007) An expert system based methodology for automating the setup planning in computer-aided process planning for rotationally symmetrical parts. Int J Adv Manuf Syst 10:81–93
13. Dereli T, Filiz IH (1999) Optimisation of process planning functions by genetic algorithms. Comput Ind Eng 36:281–308
14. Descotte Y, Latombe JC (1981) GARI: a problem solver that plans how to machine mechanical parts. In: Proceedings of IJCAI-7, Vancouver, Canada, pp 766–772
15. Dixit US (2010) Application of neural networks and fuzzy sets to machining and metal forming. In: Davim JP (ed) Artificial intelligence in manufacturing research. Nova Science Publishers, USA
16. Dorigo M, Maniezzo V, Colorni A (1996) Ant system: optimization by a colony of cooperating agents. IEEE Trans Syst Man Cybern 26:29–41
17. ElMaraghy HA (1993) Evolution and future perspectives of CAPP. Ann CIRP 42:739–751
18. Eversheim PJ, Holz B (1982) Computer aided programming of NC machine tools by using the system AUTAP-NC. Ann CIRP 31:323–327

19. Eversheim W, Schneewind J (1993) Computer-aided process planning-state of the art and future development. Robot Comput Integr Manuf 10:65–70
20. Gaoliang P, Wenjian L, Xutang Z (2005) An internet-based system for setup planning in machining operations. In: Proceedings of the 10th IEEE international conference on engineering of complex computer systems, Shanghai, China, pp 245–251
21. Gaoliang P, Wenjian L, Yuru Z (2005) Intelligent setup planning in manufacturing by fuzzy set theory based approach. In: Proceedings of the 2005 IEEE, international conference on automation science and engineering, Edmonton, Canada, pp 130–135
22. Gologlu C (2004) A constraint based operation sequencing for a knowledge based process planning. J Intell Manuf 15:463–470
23. Guo YW, Mileham AR, Owen GW, Maropoulos PG, Li WD (2009) Operation sequencing optimization for five-axis prismatic parts using a particle swarm optimization approach. Proc ImechE Part B J Eng Manuf 223:485–497
24. Hazarika M, Dixit US, Deb S (2010) A setup planning methodology for prismatic parts considering fixturing aspects. Int J Adv Manuf Technol 51:1099–1199
25. Hebbal SS, Mehta NK (2008) Setup planning for machining the features of prismatic parts. Int J Prod Res 46:3241–3257
26. Huang SH, Liu Q (2000) Geometric tolerance normalization and its applications. Trans North Am Manuf Res Inst SME 28:305–310
27. Huang SH, Liu Q (2003) Rigorous application of tolerance analysis in setup planning. Int J Adv Manuf Technol 3:196–207
28. Huang SH, Xu N (2003) Automatic set-up planning for metal cutting: an integrated methodology. Int J Prod Res 41:4339–4356
29. Jang JR, Sun C, Mizutani E (1997) Neuro-fuzzy and soft computing: a computational approach to learning and machine intelligence. Prentice-Hall, Englewood Cliffs, pp 7–9
30. Jiang B, Baines K, Zockel W (1997) A new coding scheme for optimization of milling operations for utilization by a generative expert CAPP system. J Mater Process Technol 63:163–168
31. Jiang ZG, Zhang H, Xiao M (2008) Web-based process database support system for green manufacturing. Appl Mech Mater 10–12:94–98
32. Joshi SB, Hoberecht WC, Lee J, Wysk RA, Barrick DC (1994) Design, development and implementation of an integrated group technology and computer aided process planning system. IIE Trans 26:2–18
33. Kannan M, Saha J (2009) A feature-based generic setup planning for configuration synthesis of reconfigurable machine tools. Int J Adv Manuf Technol 43:994–1009
34. Kennedy J, Eberhart R (1995) Particle swarm optimization. In: Proceedings of the IEEE international conference on neural networks, vol IV, Perth, Australia, pp 1942–1948
35. Kim IH, Oh JS, Cho KK (1996) Computer aided setup planning for machining processes. Comput Ind Eng 31:613–617
36. Kiritsis D (1995) A review of knowledge-based expert systems for process planning. Methods and problems. Int J Adv Manuf Technol 10:240–262
37. Konar A (2000) Artificial intelligence and soft computing. CRC Press, USA
38. Krishna AG, Rao KM (2006) Optimisation of operations sequence in CAPP using an ant colony algorithm. Int J Adv Manuf Technol 29:159–164
39. Lau H, Jiang B (1998) A generic integrated system from CAD to CAPP: a neutral file-cum-GT approach. Comput Integr Manuf Syst 11:67–75
40. Lau HCW, Lee CKM, Jiang B, Hui IK, Pun KF (2005) Development of a computer-integrated system to support CAD to CAPP. Int J Adv Manuf Technol 26:1032–1042
41. Lee HC, Jhee WC, Park HS (2007) Generative CAPP through projective feature recognition. Comput Ind Eng 53:241–246
42. Leung HC (1996) Annotated bibliography on computer-aided process planning. Int J Adv Manuf Technol 12:309–329
43. Li L, Fuh JYH, Zhang YF, Nee AYC (2005) Application of genetic algorithm to computer-aided process planning in distributed manufacturing environments. Robot Comput Integr Manuf 21:568–578

44. Lin AC, Lin SY, Diganta D, Lu WF (1998) An integrated approach to determining the sequence of machining operations for prismatic parts with interacting features. J Mater Process Technol 73:234–250

45. Liu Z, Wang L (2007) Sequencing of interacting prismatic machining features for process planning. Comput Ind 58:295–303

46. Lv P, Zhu C (2005) On process planning: an ant algorithm based approach. In: Proceedings of the IEEE international conference on mechatronics and automation, Niagara Falls, Canada, pp 1412–1415

47. Mei J, Zhang HC, Oldham WJB (1995) A neural networks approach for datum selection. Comput Ind 27:53–64

48. Mohemmed AW, Sahoo NC, Geok TK (2008) Solving shortest path problem using particle swarm optimization. Appl Soft Comput 8:1643–1653

49. Nadir Y, Chaabane M, Marty C (1993) PROCODE-automated coding system in group technology for rotational parts. Comput Ind 23:39–47

50. Nikolopoulos C (1997) Expert systems: introduction to first and second generation and hybrid knowledge based systems. Marcel Dekker Inc., New York

51. Ong SK, Nee AYC (1994) Application of fuzzy set theory to setup planning. Ann CIRP 43:137–144

52. Ong SK, Nee AYC (1995) Process modelling and formulation for the development of an intelligent set-up planning system. Proc ImechE Part B J Eng Manuf 209:455–467

53. Ong SK, Nee AYC (1996) Fuzzy-set-based approach for concurrent constraint set-up planning. J Intell Manuf 7:107–120

54. Ong SK, Nee AYC (1997) Automating set-up planning in machining operations. J Mater Process Technol 63:151–156

55. Ong SK, Nee AYC (1998) A systematic approach for analyzing the fixturability of parts for machining. ASME J Manuf Sci Eng 120:401–408

56. Rameshbabu V, Shunmugam MS (2009) Hybrid feature recognition method for setup planning from STEP AP-203. Robot Comput Integr Manuf 25:393–408

57. Sadaiah M, Yadav DR, Mohanram PV, Radhakrishnan P (2002) A generative computer aided process planning system for prismatic components. Int J Adv Manuf Technol 20:709–719

58. Scallan P (2003) Process planning-the design/manufacture interface. Butterworth-Heinemann Publications, Burlington

59. Shunmugam MS, Mahesh P, Bhaskara Reddy SV (2002) A method of preliminary planning for rotational components with C-axis features using genetic algorithm. Comput Ind 48:199–217

60. Sormaz DN, Khoshnevis B (1996) Process sequencing and process clustering in process planning using state space search. J Intell Manuf 7:189–200

61. Wang L, Feng HY, Cai N (2003) Architecture design for distributed process planning. J Manuf Syst 22:99–115

62. Wong TN, Chan LCF, Lau HCW (2003) Machining process sequencing with fuzzy expert system and genetic algorithms. Eng Comput 19:191–202

63. Wu HC, Chang TC (1998) Automated setup selection in feature-based process planning. Int J Prod Res 36:695–712

64. Wu R, Zhang H (1998) Object-oriented and fuzzy-set-based approach for set-up planning. Int J Adv Manuf Technol 14:406–411

65. Wysk RA, Chang TC, Wang HP (1988) Computer integrated manufacturing software and student manual. Delmar Publishers, New York

66. Xu N, Huang SH (2006) Multiple attribute utility analysis in setup plan evaluation. ASME J Manuf Sci Eng 128:220–227

67. Xu N, Huang SH, Rong YK (2007) Automatic setup planning: current state-of-the-art and future perspective. Int J Manuf Technol Manage 11:193–207

68. Yao S, Han X, Yang Y, Rong Y, Huang SH, Yen DW, Zhang G (2007) Computer aided manufacturing planning for mass customization: part 2, automated setup planning. Int J Adv Manuf Technol 32:205–217

69. Yip-Hoi D (2002) Computer-aided process planning for machining. In: Nwokah ODI, Hurmuzlu Y (eds) The mechanical systems design handbook, modelling, measurement, and control. CRC Press, USA
70. Zadeh LA (1965) Fuzzy sets. Inf Control 8:338–353
71. Zhang HC, Lin E (1999) A hybrid graph approach for automated setup planning in CAPP. Robot Comput Integr Manuf 15:89–100
72. Zhang YF, Nee AYC, Ong SK (1995) A hybrid approach for setup planning. Int J Adv Manuf Technol 10:183–190

Chapter 4
A Fuzzy Set Based Setup Planning Methodology

Abstract In this chapter application of fuzzy set theory to setup planning is described. Fuzzy set theory can deal with the uncertainty and imprecision. At actual shop floor, one faces the problem of uncertainty, imprecision and subjectiveness. Fuzzy set theory can circumvent this problem to some extent. Two examples have been illustrated to show the efficacy of fuzzy set theory, one concerning with feature precedence relation and other with datum selection. A practical strategy for adaptively controlling the setup plan is also suggested.

Keywords Setup planning · Fuzzy set theory · Membership grade · Burr control · Datum selection · Adaptive learning

4.1 Introduction

The previous chapters are devoted to introduction of setup planning in machining context and discussions on different approaches used to solve setup planning problems. It is observed over time that the use of AI and soft computing techniques is gradually increasing for solving setup planning problems giving better results compared to traditional methods of setup planning. Some of the important soft computing techniques used by researchers are artificial neural networks, fuzzy logic, expert systems, and evolutionary algorithms. In this chapter, an example of using soft computing technique to solve setup planning problem is presented. Expert system is used to build the basic framework of the setup planning system. Moreover, fuzzy sets are used to deal with the uncertainty associated with setup planning knowledge. The basics of the fuzzy set based setup planning expert system are described in detail in the present chapter.

In this chapter, an expert system is described as a representative basic setup planning tool. An expert system is an artificial intelligence (AI) tool which emulates the problem solving logic of a human brain and arrives at a solution by reasoning capability. Three main components of an expert system are database,

© The Author(s) 2015

M. Hazarika and U.S. Dixit, *Setup Planning for Machining*,
SpringerBriefs in Manufacturing and Surface Engineering,
DOI 10.1007/978-3-319-13320-1_4

knowledge-base and inference engine (Fig. 3.2). A background on expert systems is presented in Sect. 3.2.1. Expert system gives better physical insight to a problem. Usually the rule based approach is chosen because it is easier to understand and implement. The rules are represented in natural language (e.g. English). Therefore the knowledge-base can be developed rapidly without the need to perform extensive programming. It is possible to add, delete, and modify rules in the knowledge-base without extensive system changes and recoding the program. Rule based approach is general enough to capture knowledge from different sources and simple enough to provide an easily maintainable environment. It has been implemented on a PC using the expert system shell CLIPS, an acronym for C Language Integrated Production System [1]. Given the information about different features present in a part, machining operations, machine tools, cutting tools and material properties as input, the setup planning system automatically performs the tasks of setup formation, operation sequencing, and datum selection for each setup and generation of information related to fixturing. It is capable of generating setup plans for machining prismatic parts containing different types of features, e.g. face, hole, slot, pocket, step, chamfer etc. including interacting features. Two volumetric features are defined as interacting features if their boundaries intersect, so that they share a non-empty, common volume. An introduction to expert system shell CLIPS is presented in the following section.

4.2 CLIPS: An Expert System Shell

CLIPS is an expert system shell, an acronym for C Language Integrated Production System. CLIPS was first developed in 1986 by the software technology branch, NASA, Johnson Space Centre and has been undergoing continuous refinement and improvement since then. It is designed to facilitate the development of software to model human knowledge and expertise. CLIPS provides support for rule-based, object-oriented, and procedural programming. Some important features of CLIPS are flexibility, easy integration with external systems and availability. CLIPS is now maintained independently from NASA as a public domain software. CLIPS is called an expert system tool because it provides a complete environment for developing expert systems which includes features such as an integrated editor and a debugging tool. A program written in CLIPS consists of rules and facts. The basic features of rule-based programming capabilities of CLIPS are discussed hereunder.

A set of knowledge representation structures (called Construct) are provided in CLIPS to facilitate insertion of facts and rules into the expert system. Some examples of CLIPS Construct are Deftemplates, Deffacts and Defrules. In order to solve a problem, a CLIPS program must have data or information about the problem which it can reason with. A piece of information is called a fact in CLIPS. Before facts can be entered, CLIPS must be informed of the list of valid fields/attributes associated with each fact. The construct Deftemplate is used to define the format for representation of facts. It is a list of named fields called slots used to store various attributes. For example, the facts about a machining feature may be entered in the format of a Deftemplate consisting of several slots as follows:

(Deftemplate: machining_feature
(slot number (type INTEGER))(slot name(type SYMBOL))
(slot type(type SYMBOL)) (slot subtype(type SYMBOL)))

According to the above Deftemplate, a machining feature can have the slots/attributes such as feature number, feature name, feature type and subtype. Following the above format, the facts about a group of machining features can be entered using the Deffacts construct as follows:

(Deffacts: machining_feature_list
(feature (number 1) (name FACE) (type EXTERNAL) (subtype PRIMARY))
(feature (number 2) (name HOLE) (type INTERNAL) (subtype SECONDARY))
(feature (number 3) (name STEP) (type EXTERNAL) (subtype SECONDARY)))

The Defrule construct is used to build the rule-base of the expert system. For example there is a rule: IF a hole is to be drilled on a chamfered face, THEN drill the hole first and then chamfer.

Using Defrule construct, the rule is inserted to the rule-base as follows:

(Defrule::feature_precedence_constraint
(feature(number ?a)(name HOLE)(type INTERNAL)(subtype SECONDARY)
(adjacent_features ?b)(adjacent_feature_names CHAMFER))
=> (assert(feature_precedence ?a ?b)))

Besides dealing with symbolic facts, CLIPS also can perform numeric calculations. Functions have been included in CLIPS for performing various calculations. The inference engine of CLIPS is based on the forward chaining strategy. It attempts to match the patterns (antecedent part) of rules against facts in the fact-list. If all the conditions in the antecedent part of a rule match facts, the rule is activated and put on the agenda. The agenda is a collection of activations of those rules which have found matching facts. When multiple activations are on the agenda, CLIPS automatically orders the activations on the agenda in terms of increasing priority. The priority order for firing of rules can be implemented by using a unique feature of CLIPS called 'salience'. It is a mechanism to assign priorities (in terms of numeric values) to rules when multiple rules are present. A rule with higher salience fires before a rule with lower salience. The inference engine sorts the activations according to their salience. This sorting process is called conflict resolution.

All the rules and the facts about the problem to be solved are to be encoded following the syntax of CLIPS and saved as files with the extension .clp. The program is executed using the CLIPS expert system shell version 6.3 compiler under the Windows environment. At the time of execution of the expert system program, the following steps are to be performed:

- Loading of the program files from the knowledge-base into the CLIPS environment using the option 'Load CLIPS Construct' from File menu.
- Loading of the input data about the part to be machined and machining operations from database into the CLIPS environment using the option 'Load CLIPS Construct' from File menu.
- Execution of the program by first selecting 'Reset' and then 'Run' options from the Execution menu.

4.3 Architecture of the Setup Planning Expert System

The overall architecture of the proposed setup planning expert system is shown in Fig. 4.1. Its modules are the database, the rule-based knowledge-base, a fixturing information generation module, uncertainty and feedback module, the inference engine and the user interface. It has been implemented on a PC using the expert system shell CLIPS described in the previous section. The detail of each module of the proposed system is discussed hereunder.

4.3.1 Database

The database contains the declarative knowledge that includes the detailed information about the different features present in a part and the machining operations required to produce them. This information is presented in the form of data files and given by the user as input to the expert system. Information about machine tools, cutting tools and material properties are also to be included in the database. The database also contains mathematical functions and external programs that are necessary for performing different calculations. The following section explains the format of representation of the input data to the expert system.

Fig. 4.1 Architecture of the setup planning expert system

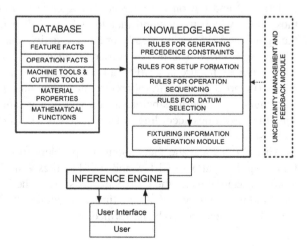

4.3.1.1 Machining Feature Information

The input information on machining features includes the type of features present in a part, their dimensions, the geometric tolerance relationship with other features, the tool access directions (TAD), feature identifier number, etc. Each of these attributes of a feature is called a 'slot'. The concept of Deftemplate (a CLIPS Construct) has been used to define the format for representation of the above input data to the expert system. Deftemplate is a list of named fields called slots used to store various above mentioned feature attributes. A slot may contain one or more fields. Each slot of a feature stores values, a single value in case of a single slot and multiple values in case of a multislot. For example, a feature can have only one name, so the attribute 'name' has single slot and has only one value. However, a feature may have more than one TAD (refer Fig. 1.7). Therefore, TAD is a multislot attribute with multiple values for TAD. The input data for a feature may be entered in the format of a Deftemplate consisting of several slots as follows:

(Deftemplate: feature
(slot number (type INTEGER))(slot name(type SYMBOL))
(slot type(type SYMBOL)) (slot subtype(type SYMBOL))
(slot length(type NUMBER)) (slot breadth(type NUMBER))
(multislot TAD (type SYMBOL))
(multislot ref-features(type INTEGER))
(multislot adjacent_ features(type INTEGER))
(multislot relation_with_ feature (type INTEGER))
(multislot tolerance (type NUMBER))

It has one slot each for the feature identifier number, name of the feature, type and sub-type of feature, feature length, breadth, TAD of feature, identifiers of the reference features, adjacent features and features with which it has tolerance relations, respective tolerances, etc. Using the above Deftemplate, the input data for a typical feature may be entered as follows:

(feature (number 10) (name FACE)(type EXTERNAL) (subtype PRIMARY)
(length 12.5)(breadth 8) (TAD TAD4 TAD6)(ref-features 14 16)
(adjacent_ features 7 9 12)(relation-with-feature 12 15) (tolerance 0.2 0.1)).

It states the fact that there is an external primary feature face with feature identifier number 10, having two TADs—TAD4 and TAD6, of length and breadth 12.5 cm and 8 cm respectively which is referenced with respect to features 14 and 16. The face has features 7, 9 and 12 as adjacent features. It has tolerance relations of value 0.2 and 0.1 mm with features 12 and 15 respectively.

4.3.1.2 Machining Operation Information

The information on machining operations includes operation identifier, operation type, and the feature on which it acts, its TAD and tolerance relationships with other features. Another Deftemplate has been used to define the format for representation of the input data about machining operations as follows:

(Deftemplate: operation
(slot number (type INTEGER)) (slot type (type SYMBOL))
(slot on-feature (type INTEGER)) (multislot TAD (type SYMBOL))
(multislot relation-with-feature (type INTEGER))
(multislot tolerance (type NUMBER))

The input data for a machining operation may be entered in the above Deftemplate format as follows:

(operation (number 100) (type mill) (on-feature 5) (TAD TAD3 TAD5)
(relation-with-feature 7 12) (tolerance 0.1 0.2)).

It states that milling operation with the identifier number 100 is to be performed on feature 5. The operation has two TADs—TAD3 and TAD5 and tolerance relations of value 0.1 and 0.2 mm with features 7 and 12 respectively. There are two ways by which the above input data on features and machining operations can be entered into the developed expert system. It can be saved as data files with the extension .clp and loaded from the file into the expert system environment at the time of execution. Alternatively, it may be also directly entered manually by typing through a user interface.

4.3.1.3 Mathematical Functions and Other Required Information

In addition to feature and machining operation information, database contains different mathematical functions to perform mathematical calculations, e.g. functions to compare the tightest tolerance relations, functions to determine the largest area face. Moreover, information on machine tools and cutting tools to machine the features present in a part, material properties of the workpiece material are to be given as input to the expert system through the database.

4.3.2 Knowledge-Base

The knowledge-base contains the problem solving knowledge of the expert system. Setup planning knowledge is incorporated in the form of IF–THEN rules in the knowledge-base. The necessary knowledge for formulating the rules is based on heuristic and expert knowledge from various sources such as handbooks, textbooks and interviews with experts and skilled machinists. Some knowledge is gathered from observations of actual machining in the shop floor. Four sets of rules have been developed for generation of machining precedence constraints, feature grouping and setup formation, machining operation sequencing within each setup and selection of datum for each setup. The detail discussions of these basic steps of setup planning are presented in Chap. 2. In the following sub-sections, only the methodology for generating the rules relevant to each step is discussed.

4.3.2.1 Generation of Machining Precedence Constraint Rules

During machining of the features comprising a part, certain precedence relations among the features are to be respected. These precedence relations arise due to basic manufacturing principles and feature interactions. An interaction between features occurs when machining of one feature affects the subsequent machining of another feature. A precedence relation between two features F1 and F2, denoted as F1 → F2, implies that F2 cannot be machined until the machining of F1 is complete. Precedence relations are discussed in detail in Sect. 2.5.1 and some of the precedence relations collected from the literature are presented in Fig. 2.3. These feature precedence relations are derived from manufacturing practice and there may be uncertainty about the validity of some assumed relations. The optimal machining sequence depends to a large extent on precedence relations. The validity of the precedence relations are to be reviewed keeping in mind the other related factors such as machining cost and time, work material properties, the required surface finish, machining passes (single or multi), etc. Details of an approach to deal with uncertainty in the precedence relations are discussed in the later part of this chapter.

In the proposed approach for setup planning, a set of rules have been developed to automatically identify the various precedence relations between the features and the machining operations needed to produce them. The following are the examples of some rules that have been included as a part of the knowledge-base for determining the precedence relations. A sample rule for generating tool interaction constraint may be written using the Defrule construct in CLIPS as follows:

(Defrule::precedence_constraint "precedence based on drilling a hole prior to the chamfer"
(feature(number ?a)(name HOLE)
(type INTERNAL)(subtype SECONDARY) (secondary_to ?b)
(adjacent_features ?b) (adjacent_features_names CHAMFER))
=>(assert(feature_precedence ?a ?b)))

The above rule states that if there is a feature *a* of name hole which is to be drilled on a chamfered face *b*, then due to tool interaction constraint, the drilling of hole *a* is to be done prior to the chamfer *b*. Another rule for generating parent-child precedence constraint is given below. It states that if there is an internal feature *a* which is nested in another feature *b*, then due to parent-child precedence constraint, the machining of feature *b* is to be done prior to the machining of *a*.

(Defrule::precedence_constraint "machining of nesting features prior to the nested features"
(feature (number ?a)(type INTERNAL)(subtype NESTED)(nested_in ?b))
=>(assert(feature_precedence ?b ?a)))

4.3.2.2 Feature Grouping and Setup Formation Rules

Setups are formed after clustering the features and their machining operations into different groups based on TAD and tolerance relations among the features. These feature clusters are assigned to different setups (refer to Sects. 2.2–2.3). An example rule for feature clustering based on tightest tolerance is as follows (Feature *a* and feature *b* are taken as examples for explaining the rules in the rule-base):

(Defrule::feature_clustering_based_on_TAD_and_tolerance
?f1<-(operation(TAD $? TAD1 $?)(on_feature ?on_feature))
?f2<-(feature(number ?on_feature))
(test(>=(length$ (fact-slot-value ?f1 tolerance))2))
(operation(TAD TAD1)(on_feature =(feature-with-tightest-tolerance ?f1)))
=>
(modify ?f1(TAD TAD1(relation_with_feature =(update-relation-with-feature ?f1))
(tolerance =(update-tolerance ?f1)(modify ?f2(TAD TAD1))

The above rule performs the following tasks:

- It identifies that there is a machining operation (fact identifier *?f1*) on feature *a* (fact identifier *?f2*) having multiple TAD and one of them is TAD1. It has tolerance relationship with more than one feature including (say) feature *b*. TAD of feature *b* is TAD1.
- Next, the mathematical function *feature-with-tightest-tolerance* compares the tolerance relations of machining operation on *a* and returns the fact identifier of feature *b* with which feature *a* has tightest tolerance. Operation on *a* is assigned the TAD of feature *b* i.e., TAD1.
- Lastly, the functions *update-relation-with-feature* and *update-tolerance* are used to update the slots *relation-with-feature* and *tolerance* respectively by removing the tolerance relationships between features that have been already considered.

If a multiple TAD feature has no tolerance relationship with other features, it is assigned the TAD of a feature cluster where there are the maximum numbers of features. A sample rule for assigning a single TAD to a feature having multiple TADs and no tolerance relation with other features is explained hereunder.

The machining operation (fact identifier *?f1*) for machining a slot (fact identifier *?f2*) has three TADs—TAD1, TAD2 and TAD3 and the slot does not have any tolerance relation with other features. Now, the machining operation and the feature slot are assigned a single TAD—TAD2 as there are maximum numbers of features in TAD2 feature cluster.

(Defrule::slot_with_multiple_TAD_no_tolerance_relation
?f1 <-(operation(TAD ?TAD1 ?TAD2 ?TAD3)(type mill)(on_feature ?on_feature))
?f2 <-(feature(number ?on_feature)(name SLOT))
=> (modify ?f1(TAD ?TAD2))
*   (modify ?f2(TAD ?TAD2))*

After forming all the feature clusters based on TAD and tolerance, setups are to be formed. Machining is performed in a vertical machining center (MC) equipped with rotary index table and automatic tool changer (ATC). It is possible to machine five faces of a cubic workpiece in these machines in a single setup. Therefore, machining of a part can be completed in two setups. The five common TAD feature clusters (TAD1, TAD2, TAD4, TAD5 and TAD6) as shown in Fig. 1.7a are grouped into one setup and the remaining common TAD feature cluster TAD3 is assigned to the other setup. Sample rules for formation of setups based on TAD feature cluster is shown hereunder.

(Defrule::formation_of_setup-1
?f1 <-(TAD1 _feature_cluster (operation_numbers $?operation_numbers))
=> (bind ?*setup-1*(fact-slot-value ?f1 operation_numbers))
    (bind ?*setup-1* (delete-member$(create$?*setup-1_feature_cluster*?*setup-1*)0)))

The above rule states that if there is a TAD1_feature_cluster (fact identifier *?f1*) with multiple operations all with TAD1, the operations are included in the Setup-1 in their proper sequence. Initially, setup-1 had no element, which is now updated to contain the operation numbers of TAD1 cluster.

4.3.2.3 Rules for Machining Operation Sequencing Within a Setup

Sequencing of machining operations within a setup is based on the machining precedence constraints generated as described in Sect. 2.5.1. Another important criterion for machining operation sequencing is to minimize tool changes by grouping the similar machining operations together. Grouping of similar machining operations can be performed by using a unique feature of CLIPS called 'salience'. It is a mechanism for assigning priority to various rules. Thus when multiple rules are present, this feature allows a rule with higher salience to fire before a rule with lower salience. First, a sequence of operations is created within a setup based on their precedence relations. This operation sequence can be modified by grouping operations of same tool together as long as the precedence relations are respected.

For machining operation sequencing within a setup, the information on preceding operation for each machining operation is required. For example, the preceding operation for machining a nested feature is machining of the nesting feature which is again preceded by machining of its reference feature. These information/facts are created by the rules for generation of precedence relations. An operation may have multiple preceding operations. First, a set of rules has been developed to find preceding operations for each machining operation. If a machining operation having no preceding operation is encountered, then it is assigned as the first operation of the respective setup. A sample rule for finding preceding operations *n2* and *n3* for an operation *n1* based on precedence relations is as follows:

(Defrule::preceding-operation
?f1<-(operation(number ?n1)(type ?type)(TAD_cluster ?TAD)(preceding_opn ?n2))
?f2<-(operation(number ?n1)(type ?type)(TAD_cluster ?TAD)(preceding_opn ?n3))
*=>(assert(operation(number ?n1)(type ?type)(TAD_cluster ?TAD) (preceding_*
    operation ?n2 ?n3))

The rule states that if there are two facts (fact identifiers *?f1 ?f2*) about an operation *n*1 having preceding operations *n*2 and *n*3, then a new fact is generated stating that operation *n*1 has *n*2 and *n*3 as preceding operations.

Next, a set of rules has been developed to determine the sequence of operations within a setup by scanning all the operation facts from the database (original ones and the new facts generated by firing of the rules) and then assigning each operation to one of the two setups in order of their precedence. The scanning of operation facts is continued until all the machining operations are assigned to one of the setups. A machining operation is assigned to a setup only if all its preceding operations have been assigned. Example of a rule for operation sequencing in Setup-1 is as follows:

(Defrule::sequence-setup-1
?f1<- (operation (number ?n1)(type mill)(preceding_operation ?n2))
?f2<- (operation (number ?n2)(setup setup-1))
=> (bind ?operation-setup-1 (fact-slot-value ?f1 number))
  (bind ?*sequence-setup-1-feature-cluster*(create$?operation-setup-1))

The rule states that if an operation *n*1 (fact identifier *?f1*) has a preceding operation *n*2 (fact identifier *?f2*), and *n*2 is assigned to setup-1, then operation *n*1 is also assigned to setup-1 in the proper sequence considering all other conditions are satisfied. Thus, using the precedence constraint information and developed rules as discussed above, a feasible sequence of machining operations within each setup is automatically generated. The machining operations are arranged in the sequential order in which they are to be performed.

4.3.2.4 Datum Selection Rules

Once the setups are formed, the setup datums are to be selected. The basis of datum selection is presented in detail in Sect. 2.4. Selection of proper datum is very important for tolerance requirements and functionality of the part. A sample rule for selecting primary datum based on tolerance relations for Setup-1 is shown below. It states that there is a machining operation (fact identifier *?f1*) to make a feature hole (fact identifier *?f2*) having TAD as TAD6. The operation has tolerance relations with more than one feature and it has the tightest tolerance relation with a feature face (fact identifier *?f3*) with TAD as TAD3. Therefore, the face (fact identifier *?f3*) is selected as primary datum for Setup-1.

(Defrule::selecting-primary-datum-for-setup-1
?f1<-(operation(on_feature ?on_feature)
?f2<-(feature(number ?on_feature)(name HOLE) (TAD TAD6))
?f3<-(feature(name FACE)(TAD TAD3))
(test (>= (length$ (fact-slot-value ?f1 tolerance)) 2))
(operation (on_feature =(feature-with-tightest-tolerance ?f3)) (TAD TAD3))
=> (assert(=PRIMARY-DATUM-SETUP-1(feature-with-tightest-tolerance ?f3))

Another priority for selecting datum is surface area of a face. A rule for selecting datum based on maximum area face is shown below. It states that there is a machining operation (fact identifier *?f1*) on a feature face (fact identifier *?f2*) having TAD as TAD6. A function *'face-having-largest-area-for-primary-datum'* compares the areas of the different faces and returns the fact identifier of the face having maximum area. The identified face has the largest area among the candidate faces for datum and it is selected as primary datum for Setup-1.

(Defrule::selecting-primary-datum-for-setup-2
?f1<-(operation(on_feature ?number)(TAD TAD6))
?f2<-(feature(name FACE)(number ?number)(TAD TAD6))
=> (if (= ?number ?*face-having-largest-area-for-primary-datum*)
  then (duplicate ?f2 (name PRIMARY-DATUM-SETUP-1)))

For selecting secondary datum, all the faces perpendicular to the primary datum are considered and the largest face is selected as the secondary datum. The tertiary datum is the largest face which is perpendicular to both primary and secondary datum.

All the above knowledge-base rules are coded using the language format of the CLIPS expert system shell and saved in the knowledge-base as files with the extension .clp. At the time of execution of the expert system program, the rules are to be loaded from the knowledge-base files into the expert system environment. The modular nature of the proposed expert system makes it easier to incorporate knowledge and expand the knowledge-base by incremental development. The rules are easily understandable and editable by the user.

4.3.3 The Inference Engine

The inference engine contains and realizes the decision making strategy. It is that part of the CLIPS expert system shell which is already programmed and ready for use. The inference engine is separated from the knowledge-base and is an independent module that makes the expert system more flexible. The inference mechanism in CLIPS expert system shell is based on forward chaining strategy where a line of reasoning is formed by chaining the IF-THEN rules in the knowledge-base to arrive at a decision. The feature and operation facts of a part to be machined are stored as data files and loaded into the CLIPS environment along with the rules from the knowledge-base module. The inference engine draws inference by deciding which rules are satisfied by facts, assign priority to the rules, and execute the rules with the highest priority.

4.3.4 The User Interface

The user interface is that part of the CLIPS expert system shell which provides the mechanism by which the user interacts with the expert system. It provides an easy access for the user to the expert system through a communication interface. The information on the various attributes of the features and the machining operations of a component, information on machine tools and cutting tools, and material properties of the workpiece materials are to be given as input by the user. Additional information and some required on-the-spot information on the shop floor can also be provided interactively. The user interface allows the user to perform various tasks, such as creating and editing the database and knowledge-base files using a text editor, saving the text files, loading the saved files into CLIPS environment, and executing the expert system program. It also provides commands for viewing the current state of the system, keeping track on the steps of the execution, recording the number of rules fired, time taken and so on. Output of the final results and decisions are also communicated to the user through this interface. A simple explanation facility is also provided in the CLIPS expert system shell. The explanation facility displays the reason behind firing a certain rule and is helpful in debugging the program. For example, it displays all the facts from the database which satisfies the IF part of a certain rule from the knowledge-base. It is helpful in debugging the knowledge-base of the expert system.

By following the methodology presented above, Hazarika et al. [2] used the CLIPS expert system shell to develop a setup planning expert system for machining of prismatic parts. The expert system is capable of automatically performing different setup planning tasks. After development of the expert system program, it is important to validate its performance on different parts. Accordingly, the performance of the developed setup planning expert system is validated on a variety of parts. Uncertainty management in the setup planning knowledge by fuzzy sets is discussed in the following subsection. A detailed discussion on setup planning with fixturing information is presented in Chap. 6.

4.4 Application of Fuzzy Set for Uncertainty Management

Uncertainty of the acquired knowledge is an important factor in the knowledge-based systems which affect the performance of the systems. In knowledge-based systems, experts' knowledge is acquired from different sources and represented as domain knowledge in the form of IF-THEN rules. However, incomplete information, imprecision and vagueness in the acquired knowledge lead to uncertainty. The performance of knowledge-based systems depends to a large extent on the way uncertainty is managed by the system. Considering the fact that setup planning has to be carried out in an environment of uncertainty, a fuzzy set based approach is used to deal with uncertainty in this work. Uncertainty in the knowledge, particularly in feature precedence relations and datum selection are considered. Moreover, a fuzzy set based strategy for adaptive learning from the feedback received from actual production stage is proposed. The proposed setup planning system can modify and adapt the knowledge-base with the help of the feedback received from the actual machining conditions on the shop floor. Traditional software systems used for automating setup planning are static in nature and they do not respond to the changes in the situation. It is necessary and important for such experts systems to keep learning and evolving from experience. For better understanding of application of fuzzy sets, a background of fuzzy sets is presented in Sect. 3.2.2.1. Subsequent discussion of this section is based on Hazarika et al. [2].

4.4.1 Uncertainty in the Feature Precedence Relations

Feature precedence relations arise from basic manufacturing principles and feature interactions. Different precedence relations are obtained due to area/volume feature interactions, tolerance relations, feature accessibility, tool interaction, fixturing interaction, datum/reference/locating requirements, and constraint of good manufacturing practice. The optimal machining sequence depends on the precedence relations to a large extent. Some precedence relations such as 'datum features, reference features, and parent features are to be machined first' have no uncertainty in them. They form the basis for definite rules. However, there may be uncertainty about some assumed precedence relations. The validity of these precedence relations are to be reviewed keeping in mind the other related factors such as machining cost and time, work material properties, the required surface finish, etc. Figure 4.2 shows some uncertain precedence relations.

Figure 4.2a shows the precedence relation for drilling two concentric holes of different diameters and depth. Applying good manufacturing practice, drilling of smaller depth hole precedes the longer depth hole. However, the precedence relation is not certain as the decision depends on many related factors like hole dimensions, ease of access, tool used, possibility of tool damage, material properties, cutting parameters, etc. In some cases drilling followed by counter boring is preferred. Similarly, Figs. 4.2b–d show the precedence of machining the faces first and then drilling/chamfering/machining pocket. However, some ambiguity arises in this type of relations. Some similar type of uncertain relations may be 'machine the face first and then machine a slot/step/pocket on the face'. As some material is removed in the shape of a slot/step/pocket, it is not economical to go for machining the whole face. Decision is based on fuzzy knowledge. The surface finish required, cost and time of machining, work material property, if burr formation is there during drilling and milling, process parameters, tool geometry and tool changes are some of the related factors that affect the decision. The certainty of these precedence relations is to be reviewed keeping in mind the related factors. As optimal machining sequence depends on precedence relations, an approach for evaluation of these relations is very important. In the proposed approach for setup planning, the validity of the precedence relations can be evaluated with a fuzzy set based method.

The precedence relation shown in Fig. 4.2b, 'face and then drill hole' is taken as example to explain the fuzzy set based methodology for uncertainty management. The relation 'face and then drill hole' is based on uncertain knowledge and leads

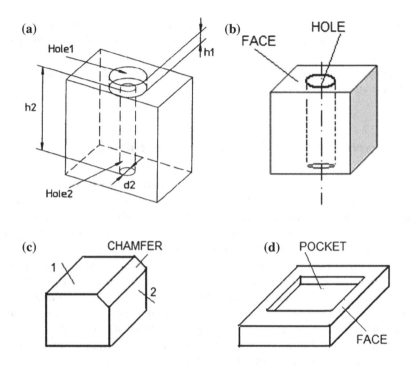

Fig. 4.2 Uncertain feature precedence relations. **a** Hole1 → Hole2. **b** Face → Drill hole. **c** Faces 1 and 2 → Chamfer. **d** Face → Pocket. **a** Presented with kind permission from Zhang et al. [4], Copyright [1995] Springer

to ambiguity. If there is burr formation during drilling, then it would be better to drill hole first and then face. Burrs are produced during drilling on both entry and exit surfaces of the workpiece due to plastic deformation of the workpiece material. Burrs are unwanted elements and burr removal involves extra cost. However, there is a possibility of burrs of negligible size which can be accepted, and facing can precede drilling in such cases. The decision depends on many other factors like work material property (ductile or brittle), the required surface finish, process parameters, tool used, total machining time and cost, etc. Significant amount of research has been devoted towards prediction and control of burr formation in drilling. It is evident from the literature that various parameters affecting burr formation in drilling are material properties, process parameters and drill geometry. Feed rate is found to be a significant factor for burr formation in these studies. Burr size is highly affected by the ductility of the workpiece material. It is observed that burr height increases with increasing ductility. Drill geometry also has a significant effect on burr shape and size. Optimization of drill geometry can minimize burr size. From the review of literature on drilling burr formation, it is observed that ductility, feed rate and tool geometry are three significant parameters which affect burr formation in drilling. In the present example, burr height is considered as a function of these three parameters. The following fuzzy rules can be formulated regarding burr formation in drilling:

IF workpiece material is ductile THEN burr height is more.
IF drill point angle is 118° THEN burr height is more.
IF feed rate is high THEN burr height is more.

Here ductility, tool geometry and feed rate are the fuzzy input parameters affecting the burr height in drilling. These input parameters may have varying effect on the output, i.e. burr height. In case of burr formation in drilling, the ductility of the workpiece material plays a dominant role compared to the other two parameters. To take into account the varying effect of the input parameters, the following relation may be adopted. If μ_{duc}, μ_{tool} and $\mu_{feedrate}$ are the individual membership grades (ranging from 0 to 1) assigned to ductility of the workpiece material, tool geometry and feed rate respectively, then it is proposed to calculate the overall computed membership grade μ_c for burr height as

$$\mu_c = \mu_{duc} \left\{ \left(\frac{\mu_{tool} + \mu_{feedrate}}{2} \right) \wedge 1 \right\}, \tag{4.1}$$

which asserts the greater effect of ductility and combined additive effect of tool geometry and feed rate on burr height in drilling. Note that $(a \wedge b)$ indicates the minimum of a and b. The ambiguous precedence rules can be written as:

Rule 1: Drill hole → Face
Rule 2: Face → Drill hole

The decision to follow Rule 1 or Rule 2 depends on many factors. The following method can be adopted for different situations.

- If for a particular combination of workpiece material, tool and feed rate, the value of μ_c is low, there will be no burr formation during drilling and therefore no precedence relation will be required. Any one of the two rules can be followed.
- If the value of μ_c is high, there will be greater chance of burr formation of considerable size and therefore precedence relation will be required. In such cases, selection of Rule 1 or Rule 2 will depend on the cost factor. There can be two options: machining without burr (following Rule 1) or machining with burrs (following Rule 2) followed by deburring techniques to remove burrs. The rule providing lower cost will be chosen.
- If the value of μ_c is neither high nor low, then the decision to follow Rule 1 or Rule 2 largely depends on the feedback received from the shop floor during actual machining stage. Certain performance measures such as presence of burrs, number of tool changes, surface finish, and total machining time are to be monitored on the shop floor at the time of actual machining and decision is to be taken based on these feedbacks. This can be done by the quality control or inspection engineer on the shop floor.

Table 4.1 Membership grades for ductility, tool geometry and feed rate

Material	μ_{duc}	Availability of drill	μ_{tool}	Feed rate (mm/min)	$\mu_{feedrate}$
Ductile like aluminum	0.9	118° point angle conventional drill	0.9	Very low 50–100	0.2
Low carbon steel like mild steel	0.7	130° point angle conventional drill	0.8	Low 100–150	0.3
High carbon steel	0.4	130° step angle step drill	0.4	Medium 150–200	0.5
Alloy steel	0.3	75° step angle step drill	0.3	High 200–250	0.7
Brittle like cast iron	0.2	60° step angle step drill	0.2	Very high 250–300	0.8

The fuzzy input parameters ductility, tool geometry and feed rate are assigned fuzzy membership grades μ_{duc}, μ_{tool} and $\mu_{feedrate}$ as given in Table 4.1. Workpiece material is graded based on their ductility. Tool geometry is graded based on the availability of the proper drill. An experimental study conducted by Ko et al. [3] showed that conventional drills with 118° drill point angle produce higher size burrs compared to 130° drill point angle and step drill with an optimum step angle of 75° (or lower) produce smaller size burrs compared to conventional drills. Feed rate is graded based on the range of feed rate considered (50–300 mm/min). Expert's opinion is sought to assign the membership grades of these input parameters.

Overall membership grade for burr height in drilling is calculated from Eq. (4.1). There may be a number of cases of different combinations of workpiece material, tool and feed rate used. Some hypothetical cases are discussed here to explain the strategy for uncertainty management.

Case 1: If the material is aluminium, tool used is 118° conventional drill and feed rate is very high, $\mu_c = 0.9\left\{\left(\frac{0.9+0.8}{2}\right) \wedge 1\right\} = 0.77$, indicating that there is great chance of burr formation. Drilling should be done prior to facing for a better surface finish.

Case 2: If the material is low carbon steel, tool used is 130° step drill and feed rate is medium, $\mu_c = 0.7\left\{\left(\frac{0.4+0.5}{2}\right) \wedge 1\right\} = 0.32$, indicating that chance of burr formation is low. Therefore, facing can precede drilling and Rule 2 can be followed. However, actual drilling process should be monitored on the shop floor to check for burr formation.

Case 3: If the material is low carbon steel, tool used is 130° point angle conventional drill and feed rate is medium, $\mu_c = 0.7\left\{\left(\frac{0.8+0.5}{2}\right) \wedge 1\right\} = 0.45$. As the value of μ_c is neither low nor high, the decision to use Rule 1 or Rule 2 depends on the feedback received from the shop floor.

Case 4: If the material is high carbon steel, tool used is 75° step angle step drill and feed rate is low, $\mu_c = 0.4\left\{\left(\frac{0.3+0.3}{2}\right) \wedge 1\right\} = 0.12$, indicating that there is negligible amount of burr formation. Therefore facing can precede drilling.

If the value of μ_c is small (<0.5), chances of burr formation is less and Rule 2 can be followed subject to shop floor feedback in the marginal cases. However, for higher values of μ_c (>0.8), there is a greater chance of burr formation of considerable size and the associated cost of machining is to be considered. The rule providing lower cost will be selected.

4.4.2 Uncertainty in the Datum Selection

The decision on selecting suitable datum for each setup depends on various factors like feature tolerance relationships, surface area of a face, its orientation, symmetry, and surface quality. Different researchers have given importance to different criteria for datum selection. As choosing the proper criteria for selecting datum is based on uncertain knowledge, fuzzy set is used to deal with the uncertainty associated with datum selection in the proposed approach. To select datum for a setup, all the faces of the part are identified. The faces having an orientation different from the faces being machined in that setup are sorted out. Criteria considered for datum selection are tolerance relations with other features, area of the candidate face and the surface quality of the face. The criterion tolerance relation means the number of tolerance relations a face has with other features and the type i.e., critical/tightest tolerance relation. However, individual objectives of these criteria are conflicting and incommensurable. Each objective is to be satisfied to a certain minimum level. Fuzzy set operations are very useful in decision making in cases where conflicting and incommensurable objectives are to be satisfied. The strategy adopted to achieve the overall objective is as follows. If μ_{gt}, μ_{gsa}, and μ_{gsq} are the individual fuzzy membership grades for good tolerance relation, good surface area and good surface quality for a particular candidate face for datum, then the overall membership grade for the face is given by

$$\mu_{O_{datum}} = \min(\mu_{gt}, \mu_{gsa}, \mu_{gsq}) \tag{4.2}$$

The overall membership grade $\mu_{O_{datum}}$ is dependent on the most poorly performing criteria for datum selection. The value of $\mu_{O_{datum}}$ for each candidate face is calculated. The face having maximum value of $\mu_{O_{datum}}$ will be selected as the primary datum for fulfilling the need for good tolerance relation, good surface area and good surface quality.

Requirements for selecting datum may vary. One requirement may be 'very good tolerance relation', 'good surface area' and 'good surface quality'. In such cases, linguistic information can be easily incorporated using fuzzy set quantifiers called hedges. Hedges are the terms that modify the meaning of a fuzzy variable. Linguistic variables and hedges are discussed in Sect. 3.2.2.3. Some examples of hedges are *very, extremely, fair, indeed,* etc. For example, if a face has membership grade μ_{gt} in the set of 'good tolerance relation', it can be assigned a membership grade μ_{gt}^2 in the set of 'very good tolerance relation'. The fuzzy membership grade for the fuzzy variable good tolerance relation is concentrated using the hedge *very* as,

$\mu_{vgt}{}^{very} = \mu^2{}_{gt}$, where μ_{vgt} is the membership grade for very good tolerance relation. After finding 'min$(\mu_{vgt}, \mu_{gsa}, \mu_{gsq})$' for all candidate faces, the face with the
maximum membership grade among these minimums is selected as datum. If the
requirement is to select very good tolerance relation, good surface area and fair surface quality, the fuzzy membership grade for good surface quality is dilated using
the hedge *fair* as, $\mu_{fsq}{}^{fair} = \sqrt{\mu_{gsq}}$, where μ_{fsq} is the membership grade for fair surface quality and datum is selected in a similar way as described above.

4.4.3 Strategy for Adaptive Learning and Updating the Knowledge-Base

Traditional software packages for automating setup planning are static in nature
and they do not respond to the changes in the situation. Capability of learning from the feedback and adaptability to the actual condition on the shop floor
is important for a setup planning system. In this section, a strategy for adaptive
learning from the feedback received from actual production stage is described.
The proposed approach enables the setup planning system to modify and adapt
the knowledge-base to the actual situations on the shop floor. Figure 4.3 shows the
flow chart for the feedback system for adaptive learning.

Checking the status of the actual machining of the parts is essential for feeding
real time information to the setup planning system. The proposed setup planning

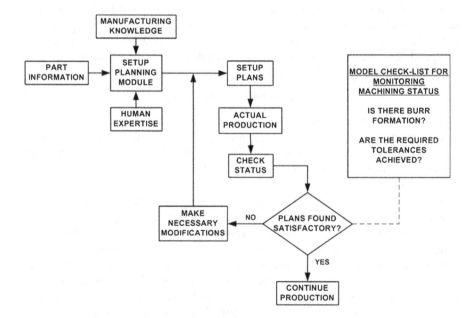

Fig. 4.3 Flow chart for adaptive learning from shop floor feedback. From Hazarika et al. [2]

system provides a checklist with the setup plan to monitor certain situations where decision is to be taken based on the shop floor feedback. Monitoring the machining conditions against the checklist is done to confirm the satisfactory performance of the setup plan. The quality control engineer will monitor if burrs are present and if desired tolerances and required surface finish are achieved. If some anomaly is found, the information is fed back to the setup planning system and modifications are made. For example, the ambiguous precedence rule, Face → Drill hole/ Drill hole → Face (as discussed in Sect. 4.4.1) is considered. For a particular combination of material, tool and feed rate for drilling a hole, Rule 2 (Face → Drill hole) is selected during setup planning stage assuming no burr will be formed. However, during actual drilling, the process is observed against the conditions given in the checklist in the setup plan and found that there is burr of considerable size. This information is to be fed back to the setup planning system so that the necessary modifications can be made to the knowledge-base and the new knowledge is stored. For certain cases when the overall membership grade μ_c for burr formation is neither high nor low (say $0.45 < \mu_c < 0.55$), decision to follow the rule, *Face → Drill hole/Drill hole → Face* depends on the feedback from the shop floor. Similarly, actual machining can be monitored in the shop floor in case of other ambiguous cases and feedback can be provided.

The strategy can be implemented as follows. Assume that the value of overall membership grade μ_c for burr formation (from Eq. 4.1) is μ_1 (which is low) for a particular combination of material, tool and feed rate indicating no burr will be formed. During actual drilling process, it is found that there is burr formation of considerable size and observed value of membership grade is assigned μ_2. Then the new value of μ_c in the knowledge-base becomes $\mu_c = (\mu_1\mu_2)^{1/2}$. Supposing, the next observed value of membership grade is μ_3, then $\mu_c = (\mu_1\mu_2\mu_3)^{1/3}$. Individual membership grades for ductility/tool geometry/feed rate can also be modified if the exact cause for burr formation can be identified. Thus the setup planning expert system keeps learning from its experience from the actual shop floor conditions and keeps updating its knowledge-base. It is necessary and important for such experts systems to keep learning and evolving from experience.

4.5 Application of the Developed Methodology

In this section, one example is presented to demonstrate the application of the fuzzy set based setup planning methodology equipped with uncertainty management strategy. The example part shown in Fig. 2.1, is considered for demonstration. It shows a component to be machined along with the detailed information on its features, dimensions, machining operations needed, TAD and tolerances among the features. Given the information about different features present in a part, machining operations, machine tools, cutting tools and material properties as input, the setup planning system automatically performs the tasks of setup formation, operation sequencing, datum selection and generation of information related

to fixturing. The fixturing consideration in setup planning is discussed in detail in
Chap. 6. The proposed setup planning expert system has been implemented on a
PC using the expert system shell CLIPS, an acronym for C Language Integrated
Production System [1]. The machining is performed in a vertical machining center
(MC) equipped with rotary index table and automatic tool changer (ATC) where
various milling as well as drilling operations can be performed. The machining
center contains simultaneously controlled three Cartesian axes X, Y, and Z. It is
possible to machine five faces of a cubic workpiece in these types of machines
in a single setup. The raw stock in Fig. 2.1 is a prismatic block of dimensions
$70 \times 60 \times 50$ mm^3 and the workpiece material is AISI 1018 steel. Taking the density of AISI 1018 steel as 7.87 g/cc, the weight of the workpiece is found to be
16.5 N. It is assumed that all the six faces (faces 1, 2, 9, 10, 11, and 12) of the
prismatic block are rough machined and only faces 1 and 2 (primary features for
the other secondary features) are considered as machining features. The through
hole 8 has parallelism tolerance 0.15 mm with the blind hole 7 and perpendicularity tolerance 0.20 mm with face 2, so it has a tighter tolerance relation with 7.
Face 1 has parallelism tolerance 0.20 mm with face 2. Face 2 also has positional
tolerance relations with features 4, 5, and 6. The uncertainty management strategy
for feature precedence and datum selection for the example part is demonstrated
hereunder.

The decision on whether to drill the through hole 8 first and then machine
face 1 or the reverse is based on uncertain knowledge. There may be drilling
burr formation, which mainly depends on ductility, tool geometry and feed rate.
Therefore, the strategy developed for uncertainty management in the feature precedence relations described in Sect. 4.4.1 is adopted. Fuzzy membership grades
μ_{duc}, μ_{tool} and $\mu_{feedrate}$ for fuzzy input parameters ductility, tool geometry and
feed rate are given in Table 4.1. Overall membership grade for burr height in drilling is calculated using Eq. (4.1). A two flute high-speed steel conventional drill
with 5 mm diameter (130° point angle and 30° helix angle) is used for drilling
hole 8, and feed rate is medium (200 mm/min) which gives the value of overall
membership grade for burr height, $\mu_c = 0.45$. As the value of μ_c is neither very
low nor very high, the decision to drill hole 8 or machine face 1 first depends on
the feedback received from the shop floor. Actual drilling process should be monitored on the shop floor to check for burr formation. This monitoring task is kept in
the check list to be provided with the setup plan.

To deal with the uncertainty of selection of datum of the example part shown
in Fig. 2.1, the six enveloping faces of the part are considered. The main parameters considered for datum selection are tolerance relation with other features, surface area and surface quality as discussed in Sect. 4.4.2. Face 2 has the maximum
number of tolerance relations with other features. It has parallelism tolerance with
feature 1, perpendicularity tolerance with feature 8, and positional tolerances with
features 4, 5, and 6. It also has more surface area (70×60 mm^2) compared to
faces 9 and 10 (60×50 mm^2 each), and 11 and 12 (70×50 mm^2 each). Table 4.2
shows the fuzzy membership grades assigned for the parameters for datum selection for the six faces of the example part. Using Eq. (4.2), the value of overall

Table 4.2 Fuzzy membership grades for the parameters for datum selection [2]

Faces	Good tolerance relation μ_{gt}	Good surface area μ_{gsa}	Good surface quality μ_{gsq}	$\mu_{O_{datum}}$
1	0.7	0.8	0.8	0.7
2	0.9	0.8	0.8	0.8
9	0.4	0.6	0.5	0.4
10	0.5	0.6	0.5	0.5
11	0.2	0.7	0.5	0.2
12	0.2	0.7	0.5	0.2

Table 4.3 Feature information of the example part

Feature	Type	Dimensions (mm)	Cutter	Cutter diameter D (mm)
1	Face	70×60	End mill	20
2	Face	70×60	End mill	20
3	Slot	$20 \times 5 \times 60$	End mill	20
4	Step	$20 \times 5 \times 60$	End mill	20
5	Step	$15 \times 5 \times 60$	End mill	20
6	Chamfer	$45 \times 10 \times 60$	Chamfer mill	16
7	Blind hole	$\phi 5 \times 20$	Drill	5
8	Through hole	$\phi 5 \times 50$	Drill	5

membership grade $\mu_{O_{datum}}$ is calculated for all the faces. From Table 4.2, the maximum value of $\mu_{O_{datum}}$ is 0.8 and the corresponding face 2 is selected as datum for machining the features 1, 3, 4, 6, 7, and 8 in one setup. This selection will ensure good tolerance, good surface area and good surface quality. Similarly, face 1 is selected as datum for machining the features 2, and 5. Thus fuzzy sets can be used to deal with the uncertainty of datum selection for setup planning.

The data file containing the information on the features, machining operations, machines and tools for the example part is loaded into the CLIPS environment along with the knowledge-based rules. The developed expert system generates the number of setups, sequence of operations within a setup, datum for each setup along with the checklist. The features of the example part and their relevant information are given in Table 4.3.

When the expert system program is executed, the final setup plan is generated automatically. It contains the number of setups, sequence of operations within a setup, datum for each setup along with the checklist. All these information for machining the example part are given in Tables 4.4 and 4.5. It requires two setups, Setup-1 and Setup-2 to machine the example part. Through hole 8 has two TADs and it is assigned TAD6 based on its tighter tolerance relation with feature 7. Features 1, 3, 4 and 6 have multiple TADs and they are assigned to TAD6 feature cluster where there is the maximum number of features.

Table 4.4 Setup plan to machine the example part

Setups	Sequence of machining operations	Primary datum	Secondary datum	Tertiary datum
Setup-1	Operation 102 on feature 2 (milling)	1	11/12	9/10
	Operation 502 on feature 5 (milling)			
Setup-2	Operation 101 on feature 1 (milling)	2	11/12	9/10
	Operation 501 on feature 4 (milling)			
	Operation 201 on feature 3 (milling)			
	Operation 301 on feature 7 (drilling)			
	Operation 302 on feature 8 (drilling)			
	Operation 400 on feature 6 (chamfering)			

Table 4.5 The checklist given with the setup plan

Checklist
(1) Check the burr height during drilling operation 302. If burr height <0.2 mm, milling face 1 precedes drilling hole 8. If burr height >0.2 mm, drilling hole 8 precedes milling face 1
(2) Check that the parallelism tolerance 0.20 mm between faces 1 and 2 is achieved. If the desired tolerance is not achieved with the ideal machining conditions, then there is a need to check for the machine capability or some inherent error present in the machine

4.6 Conclusion

In this chapter, an example of using soft computing techniques, viz. expert system and fuzzy sets to solve setup planning problem is presented. An overview of the proposed setup planning expert system is discussed. CLIPS, an expert system shell is used to develop the setup planning system in the present work. CLIPS contains a built-in inference engine, a user interface, a set of knowledge representation structures and facilities to interface with external systems. Next, different modules of the setup planning expert system are described. It contains a database, a knowledge-base, a fixturing information generation module, a module for uncertainty and feedback, inference engine and a user interface. The development steps of the database and the knowledge-base are discussed thoroughly with examples. As setup planning has to be carried out in an environment of uncertainty particularly in feature precedence relations and datum selection, fuzzy set theory is used to deal with these uncertainties. Moreover, the proposed setup planning system has the capability for modifying and adapting itself to the changes based on the shop floor data. It provides a checklist with the setup information to be used in the shop floor for providing the appropriate feedback. The feedback is used for modifying the setup plan appropriately. Finally, an example part is used to show the application of the setup planning methodology and the strategy for uncertainty management in feature precedence and datum selection.

Acknowledgements   Significant part of this chapter has been taken from Hazarika et al. [2]. We are thankful to Sage Publishers for allowing us to retain the copyright of the article.

References

1. Giarratano JC, Riley G (1998) Expert systems: principles and programming. PWS Publishers Co., Boston
2. Hazarika M, Deb S, Dixit US, Davim JP (2011) A fuzzy set based setup planning system with ability for online learning. Proc IMechE Part B J Eng Manuf 225:247–263
3. Ko SL, Chang JE, Yang GE (2003) Burr minimizing schemes in drilling. J Mater Process Technol 140:237–242
4. Zhang YF, Nee AYC, Ong SK (1995) A hybrid approach for setup planning. Int J Adv Manuf Technol 10:183–190

Chapter 5
Assigning and Fine Tuning of Fuzzy Membership Grades

Abstract Assignment of membership grades is an important task in the application of fuzzy set theory to setup planning area. There are various methods of assigning membership grades. Some of them are direct rating, polling, reverse rating, interval estimation, membership function exemplification and pair-wise comparison. There are various popular forms of membership functions like triangular, trapezoidal, Gaussian, Sigmoid, S-shaped and Π functions. The membership grades assigned by the expert can be fine tuned by the experts. In this chapter a method for fine tuning the membership grades assigned by experts is described. The method gives importance to the opinion of experts as well as relies on practical data. This method is applied to the estimation of burr size in the drilling process.

Keywords Membership grades · Membership function · Burr · Drilling · Optimization · Fuzzy set

5.1 Introduction

In the previous chapter, an example of using soft computing technique to solve setup planning problem is presented. Expert system is used to build the basic framework of the setup planning system and fuzzy sets are used to deal with the uncertainty associated with setup planning knowledge. Fuzzy set theory has been used in setup planning as evident from the literature. For a fuzzy input or output variable, membership functions/membership grades are assigned to map numeric data to linguistic fuzzy terms. However, design of membership functions/membership grades for a fuzzy set based inference system is an important issue. It greatly affects a fuzzy set based system. The problem of finding appropriate membership functions/membership grades for the fuzzy variables poses a challenge to the researchers. The competency of human experts plays a vital role in assigning these membership grades. Most of the time, different estimates of a fuzzy variable are decided based on expert's opinion. However, there is a need to optimize these estimates to enhance performance. The purpose of this chapter is to discuss

© The Author(s) 2015

M. Hazarika and U.S. Dixit, *Setup Planning for Machining*,
SpringerBriefs in Manufacturing and Surface Engineering,
DOI 10.1007/978-3-319-13320-1_5

the methods of assigning and fine tuning the membership grades. Initial membership grades are usually assigned by an expert for a fuzzy variable. Fine tuning of the expert's initial estimates of the membership grades enhances the system performance.

Different methods have been proposed in the literature for automatic generation of membership grades/membership functions for fuzzy set based inference systems. Some methods eliminate the need for an expert's opinion and knowledge is acquired from training examples. Example based learning strategy have been used in the literature to decide the membership functions. The neural networks and evolutionary algorithms also have been used for generation and optimization of membership functions. However, the common view is that there is no single best method that can be used for all applications. Choice of a method depends on the problem at hand. From the review of literature, it is evident that membership grade/membership function generation has received significant research attention over the years. However, there are limited attempts on developing a strategy that combines the best of an expert's knowledge and available data for a better solution. The experience and knowledge of an expert is valuable for initial estimates of a fuzzy parameter, although expert's knowledge may not be fully accurate. Therefore, a fine tuning strategy may be applied to the initial membership grades for finding the optimal membership grades.

5.2 Methods to Assign Fuzzy Membership Grades

How to assign membership grades to a fuzzy variable is a big challenge. It is a problem of assigning numbers to linguistic terms and there may not be a unique answer to it. For example, it is known that the surface finish produced by milling process is reasonably good and the surface finish produced by grinding process is very good. The question is what membership grade should be assigned to milled and ground surfaces in the set of "Smooth Surface". One expert may assign a membership grade of 0.7 to milled surface and 0.9 to ground surface. Another expert may assign a membership grade of 0.75 to milled surface and 0.88 to ground surface. Although both the experts assign a lower membership grade to milled surface in comparison to ground surface, their numerical values differ. Thus, membership grades are subjective, but luckily they are not arbitrary. One does not expect that any expert will assign a membership grade of 0.9 to milled surface and 0.7 to ground surface in the fuzzy set of "Smooth Surface".

The problem of subjectivity is always present in decision making field. In multi-objective optimization problems also, often the assignment of weights for various objectives differs from expert to expert. The subjectivity is unavoidable in human judgment. A number of scientific methods have been developed to capture the real meaning of subjective judgment. Sanco-Royo and Verdegay [23] have reviewed a number of methods for constructing the membership grade. These are briefly described in the context of a fuzzy set of "Smooth Surface".

5.2.1 Direct Rating

Here, the expert provides a direct rating value to the element in the fuzzy set of "Smooth Surface". For example, the expert is shown a milled surface and the following question is asked: "How smooth is this surface?" Expert is allowed to choose a suitable value between 0 and 1, which becomes the membership grade.

5.2.2 Polling

In this method, a number of experts are asked the question: "Do you agree that this surface is smooth?" Each expert is allowed to answer either "yes" or "no". The membership grade is directly obtained as the proportion of positive answers over the total number of answers.

5.2.3 Reverse Rating

The subject is provided a membership grade and is asked to which surface such a membership grade would correspond in relation to fuzzy set. For example, an expert is shown 10 surfaces and may be asked which surface can correspond to membership grade of 0.7. This process can be repeated on a number of experts and a mean can be obtained.

5.2.4 Interval Estimation

Here the experts provide an interval of possible values that best describe the fuzzy set. For example, a typical instruction to expert may be like this: "Give an interval of centre line average (CLA) values of surface roughness that make a smooth surface." The opinions of various experts can be aggregated to find out the average response.

5.2.5 Membership Function Exemplification

Here, the experts are required to provide the membership grade to several discrete points. Then, the membership function can be constructed. Typical instruction to experts may be like this: "Give the degree of belongingness of surface roughness values to the set of smooth surface." This method is also referred as continuous direct rating. In this method also the response of various experts may be averaged.

Sometimes, the experts may be required to choose among various membership function shown graphically. Most popular way of constructing membership function in this category is to ask the expert the low, most likely and high estimate of a parameter in the fuzzy set. For example, the expert may provide the CLA surface values of 0.4, 0.9 and 1.6 μm as low, most likely and high estimates respectively. The low and high values can be assigned a membership grade of 0.5 and the most likely values can be assigned a membership value of 1. With the help of these three points, a triangular membership function can be constructed. This was explained in Chap. 3 also.

5.2.6 Pair-Wise Comparison

This method has roots in the work of Saaty [21, 22]. It consists of comparing the strengths by which two objects possess the quality being analyzed. Typical question may be like this: "Between these two surfaces, which surface is smoother and by how many times?" This results in a non-symmetrical reciprocal matrix of relative weights. The eigenvector corresponding to the maximum eigenvalue provides membership grades of the objects. The eigenvector needs to be normalized. If assessment of expert is fully consistent, then the maximum eigenvalue will be equal to the number of objects, n. It can be proved that for any reciprocal matrix ($a_{ij}a_{ji} = 1$) with positive entries will have the maximum eigenvalue λ_{max} greater or equal than n. A consistency index is defined as follows:

$$\text{CI} = \frac{\lambda_{max} - n}{n - 1}. \tag{5.1}$$

Ideally, this value should be zero. The greater the value of CI, the greater is the inconsistency in the rating of expert. If the inconsistency is more, the expert should be asked to rate again. The pair-wise comparison of Saaty is the part of the analytic hierarchy process (AHP) developed by him. The major drawback is that when n is high, many pair-wise comparisons are required. The method of pair-wise comparison is illustrated with the help of an example.

Example Three surfaces are generated by lapping, grinding and milling, respectively. An expert provided the relative weights of three surfaces as per the following matrix:

	Lapping	Grinding	Milling
Lapping	1	3	9
Grinding	1/3	1	3
Milling	1/9	1/3	1

In this matrix, the row corresponding to lapping provides the relative weights of lapped surface. Relative to itself its weight should be 1. That is why the first diagonal element is 1. In fact by the same logic, all diagonal elements will be 1. The element corresponding to first row and second column is 3, which means compared to ground surface the weight of lapped surface is 3. Similarly, the element corresponding to the first row and third column tells that compared to milled surface, the weight of lapped surface is 9. If weight of the lapped surface with respect to the ground surface is 3, the weight of the ground surface with respect to the lapped surface is 1/3. That is why the element corresponding to second row and first column is 1/3.

The eigenvalues corresponding to this matrix are 3, 0 and 0. The eigenvector corresponding to the largest eigenvalue is $[0.9435\ 0.3145\ 0.1048]^{T}$. By Eq. (5.1), CI is zero indicating that the expert's estimates are consistent. By scaling the eigenvector corresponding to the largest eigenvalue such that the largest element of it becomes one, the eigenvector $[1\ 0.333\ 0.111]^{T}$ is obtained. Thus, the membership grades corresponding to lapped, ground and milled surfaces are 1, 0.333 and 0.111, respectively.

5.2.7 Review of the Recent Literature

Medasani et al. [18] provided an overview of various techniques used for membership function generation for pattern recognition. The authors are of the view that each technique for membership function generation is appropriate for a particular application and selection of a technique is governed by the specific problem. Example based learning has been widely used in the literature for generation of membership functions. Furukawa and Yamakawa [8] proposed an algorithmic approach for pattern recognition of hand written characters. The algorithms use example based learning strategy. A fuzzy neuron is assigned to each class of pattern samples that are to be recognized and the membership function for these fuzzy neurons are decided from the example based learning. Hong and Lee [12] proposed a methodology for automatic generation of membership functions and fuzzy IF–THEN rules for developing a fuzzy expert system. The system learns from a set of given training examples. Triangular membership functions are used for both input and output variables. Chen and Wang [5] are of the view that for fuzzy logic based systems, the parameter identification (i.e., deciding the number of membership functions, centre, width and cross-over slope) is an important step. A hybrid learning approach using an adaptive-network-based fuzzy inference system (ANFIS) is used to optimize the fuzzy parameters for maximizing the system performance. Liu and Pedrycz [16] proposed an algorithm for building membership functions based on an axiomatic fuzzy set (AFS) theory. The neural networks have been used to generate and optimize membership functions. Yang and Bose [25] proposed a strategy to generate membership functions for pattern recognition using self-organizing feature map technique based on unsupervised learning. A comparison with histogram method, fuzzy c-means clustering method and feed forward neural network method is presented. Medaglia et al. [17] proposed a method for generation of membership functions based on Bezier curves. An expert can provide membership grades for each point in the domain and a smooth curve can be obtained by minimizing the sum of the squared errors between the fitted membership function and data. Bai and Chen [4] developed a methodology for the construction of membership functions for the grades obtained by students. The scores of the students are inferred by fuzzy reasoning based on the constructed membership function. Choi and Rhee [6] proposed three different algorithms based on heuristics, histograms and fuzzy c-means clustering for generation of interval type-2 fuzzy membership functions for pattern recognition. In interval type-2 fuzzy membership set, for each element of a universal set, a lower and upper membership grades are defined instead of just one membership grade. Evolutionary algorithms have also been used for the optimization of membership functions. Garibaldi and Ifeachor [9] developed a fuzzy expert system for umbilical cord acid-base interpretation of newborn infants. Analysis of acid-base balance in the blood of umbilical cord gives essential information on any lack of oxygen during childbirth. Opinions of several expert clinicians are sought to rank different complex cases and these rankings are used to train the fuzzy expert system. For optimization purpose, a hybrid of simulated annealing and simplex method are used. Arslan and Kaya [2] used genetic algorithm to optimize the shape of the membership functions where the initial shape and the parameters of the membership functions are predefined. The base lengths of the input and output fuzzy variables are adjusted to find the optimal membership functions. Bagis [3] proposed an approach for attaining optimum membership functions for a fuzzy logic controller for the operation of spillway gates of reservoirs during floods. Membership functions for the input and output fuzzy variables are first selected based on experience and intuition. Optimal membership functions are obtained with Tabu search algorithm that provides a better performance of the controller.

5.3 Some Popular Forms of Membership Functions

In Chap. 3, a linear triangular membership function was introduced. Figure 5.1 shows a trapezoidal membership function. This membership function is similar to linear triangular membership function. The difference is that in the case of trapezoidal membership function, for an interval of parameter values, the membership grade is 1. This type of membership function is constructed when the experts provide the most likely estimate as an interval number. In general, linear membership functions are preferred because of ease of computations. However, there are a lot of popular non-linear membership functions that are used to mimic real-life situation.

Figure 5.2 shows a Gaussian membership function. A Gaussian membership function $\mu(x)$ may be expressed as

$$\mu(x) = e^{-\frac{1}{2}\left(\frac{x-m}{\sigma}\right)^2}, \tag{5.2}$$

where m is the most likely estimate of the parameter at which the membership grade is 1 and σ controls the spread. If the probability distribution is Gaussian,

Fig. 5.1 A trapezoidal
membership function

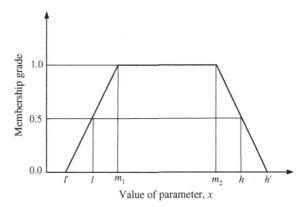

Fig. 5.2 A Gaussian
membership function

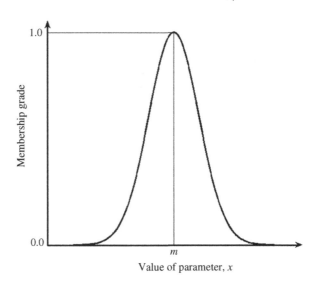

then approximately 99.73 % values lie within $\pm 3\sigma$ limit from the mean. Thus, σ may be calculated from the low and high estimates of the parameters as follows:

$$\sigma = \frac{h - l}{6}. \tag{5.3}$$

Figure 5.3 shows a Sigmoid function. The expression for the membership function is given by

$$\mu(x) = \frac{1}{1 + e^{-\alpha(x-b)}}. \tag{5.4}$$

The value of this function at $x = b$ is 0.5. This point is called crossover point, where the curvature changes sign. For large values of x, the membership grade asymptotically reaches 1 and for small values of x, it asymptotically reaches 0. A similar function called S-function (Fig. 5.4) is proposed by Zadeh [26]. It is given by

$$\mu(x) = \begin{cases} 0 & \text{if } x \le a \\ 2\left(\frac{x-a}{c-a}\right)^2 & \text{if } a < x \le b \\ 1 - 2\left(\frac{x-c}{c-a}\right)^2 & \text{if } b < x \le c \\ 1 & \text{if } x > c \end{cases}, \tag{5.5}$$

Fig. 5.3 A sigmoid membership function

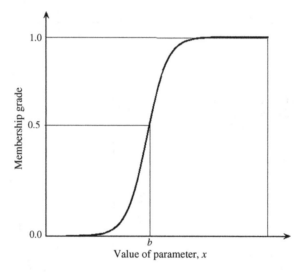

Fig. 5.4 Zadeh's S membership function

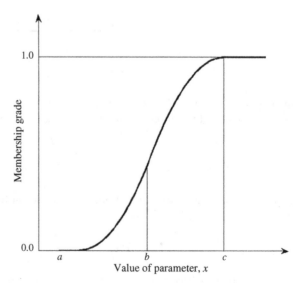

Fig. 5.5 Zadeh's Π
membership function

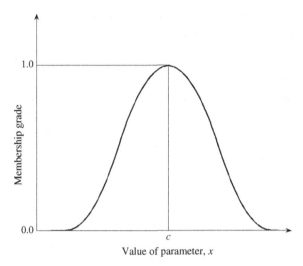

Usually, b is taken as $(a + c)/2$. In that case, at $x = b$, the membership grade is
0.5. Therefore, $x = b$ is a crossover point. For $a < x < b$,

$$\frac{d\mu}{dx} = 4\left(\frac{x - a}{c - a}\right). \tag{5.6}$$

For $b < x < c$,

$$\frac{d\mu}{dx} = 4\left(\frac{c - x}{c - a}\right). \tag{5.7}$$

Thus, before the crossover point, the rate of change of the membership function
increases from 0 to 2 and after the crossover point it keeps on decreasing from 2
to 0. Zadeh's Π function is obtained by mirroring the S-function about $x = c$ as
shown in Fig. 5.5.

5.4 Fine Tuning of Membership Grades

There are many situations where the membership grades of two or more fuzzy variables are combined to obtain an overall membership grade. For example, consider
that a certain job requires sufficient amount of intellectual ability as well as physical fitness. Now, if a candidate has a membership grade of μ_{in} in the set of 'intellectual' and a membership grade of μ_{ph} in the set of 'physical fitness', then his/her
overall membership grade μ_{c} in the set of 'suitable candidates' can be employed
using some fuzzy set theoretic operation, such as

$$\mu_{\text{c}} = \min\left(\mu_{\text{in}}, \mu_{\text{ph}}\right) \tag{5.8}$$

In general, the overall computed/predicted membership grade μ_c of a fuzzy output variable for n fuzzy input variables can be expressed as

$$\mu_c = f(\mu_1, \mu_2, \ldots, \mu_n), \tag{5.9}$$

where μ_i ($i = 1$ to n) denotes the membership grade corresponding to ith fuzzy set and f is the appropriate fuzzy set theoretic operation. The success of a fuzzy set based method depends on the accurate assignment of membership grades as well as use of an appropriate fuzzy set theoretic operation for obtaining an overall membership grade. The errors in the estimation of these quantities may reinforce or nullify one another. Hence, it may not be appropriate to apply a fuzzy set based method without the involvement of an expert. However, the estimates of experts may be fine tuned following a systematic mathematical procedure.

Hazarika et al. [10, 11] have proposed a method for fine tuning of the membership grades. It is assumed that the confidence level in the estimation of μ_c is the highest, followed by the confidence in the appropriateness of f. There may be significant uncertainty in the estimation of μ_i ($i = 1$ to n) and expert may specify it as a range, rather than a fixed real number. The task is to fine tune the values of μ_is for satisfying Eq. (5.9). In doing so, there should not be significant deviation from the opinion of the expert. If this task cannot be completed satisfactorily, then the operator f has to be modified. Compared to the fuzzy rule based inference systems, the adopted methodology is much simpler. In fuzzy rule based systems, inference procedure consists of several steps. For example, Mamdani inference process is performed in four steps: fuzzification of the input variables, rule evaluation, aggregation of the rule outputs and defuzzification for a crisp output value. However, in Eq. (5.9), the fuzzy input and output variables are related through an appropriate fuzzy set theoretic operator f and output is obtained in a single step by executing operation f on the input variables μ_i ($i = 1$ to n).

The membership grades assigned by the expert can be slightly modified based on the observed data. The difference between the computed/predicted and observed overall membership grades can be minimized in the least square sense. The overall methodology comprises the following steps:

Step 1: Data is generated from experiments/polling/interviews with experts for the fuzzy output variable for which the overall observed membership grade μ_o is to be obtained.

Step 2: Overall observed membership grade μ_o is constructed based on the data.

Step 3: Membership grades μ_i ($i = 1$ to n) for the fuzzy input variables, their variable bounds, and the appropriate fuzzy set theoretic operator f is selected based on expert's knowledge.

Step 4: Operator f is applied to μ_i ($i = 1$ to n) to obtain the value of overall computed membership grade from Eq. (5.9) which is denoted by μ_c.

Step 5: Objective is to minimize the difference between μ_c and μ_o so that observed and computed values of overall membership grades are close to each other giving a suitable solution for the membership grades μ_i ($i = 1$ to n) of the fuzzy input variables. The optimization problem is given by

$$\text{Minimize error } E = \sum_{i=1}^{k} (\mu_c - \mu_o)^2, \tag{5.10}$$

subject to constraints and variable bounds. In Eq. (5.10), k is the number of independent observation. The design variables are μ_i ($i = 1$ to n), i.e. the membership grades of individual attributes.

Step 6: For fine tuning the membership grades of fuzzy input variables, the following two criteria are considered; (i) accuracy of the solution and (ii) deviation of expert's opinion. The accuracy of the solution is expressed in the linguistic form and evaluated as explained below.

The initial estimates of membership grades for the fuzzy input variables and their variable bounds are decided by an expert. The overall computed membership grade μ_c is calculated and compared with the overall observed membership grade μ_o. The root mean square (RMS) error value is calculated as per the following equation:

$$\text{RMS error} = \sqrt{\frac{\sum_{i=1}^{k} (\mu_c - \mu_o)^2}{k}} \tag{5.11}$$

An accurate solution will have a low value of the RMS error. Table 5.1 shows the RMS errors and their equivalent numerical values. A solution for μ_i ($i = 1$ to n) is assigned a numerical value for the level of accuracy attained. A solution with very poor/poor level of accuracy is not acceptable.

Step 7: If accuracy of the solution is not excellent, the variable bounds of the μ_i ($i = 1$ to n) given by the expert are relaxed slightly and a new solution is obtained. For the new solution, each μ_i is compared with the variable bound provided by the expert and its deviation from the given bound is calculated. For a μ_i if there is no deviation of the variable bound provided by the expert, it is considered the best. Table 5.2 shows the numerical values assigned to a μ_i based on the deviation of expert's opinion.

Step 8: The new solution is also evaluated for accuracy as in Step 6. For an acceptable solution, the minimum level for accuracy as well as deviation of expert's opinion should be satisfactory. A solution with very poor/poor quality either in accuracy criterion or in deviation of expert's opinion criterion is not acceptable.

Table 5.1 The quality of solution based on the accuracy [11]

RMS error	Solution quality	Equivalent numerical value
<0.08	Excellent	10
0.08–0.1	Very good	9
0.1–0.12	Good	8
0.12–0.15	Satisfactory	7
0.15–0.17	Poor	4
>0.17	Very poor	2

Table 5.2 The level of deviations of expert's opinion [11]

Change in variable bound of a μ_i given by expert	Level of deviation	Equivalent numerical value
No change	Excellent	10
0.02	Very good	9
0.05	Good	8
0.10	Satisfactory	7
0.15	Poor	4
0.20	Very poor	2

Table 5.3 Evaluation based on accuracy and deviation of expert's opinion [11]

Acceptable solutions	Numerical value assigned for deviation of expert's opinion $\mu_1, \mu_2, \mu_3 \ldots\ldots\ldots \mu_n$	Overall quality value for a solution $(E_t = \sum e_{ij}/n)$	Numerical value for accuracy (A_t)
1	$e_{11}, e_{12}, e_{13}, \ldots\ldots, e_{1n}$	E_1	A_1
2	$e_{21}, e_{22}, e_{23}, \ldots\ldots, e_{2n}$	E_2	A_2
3	$e_{31}, e_{32}, e_{33}, \ldots\ldots, e_{3n}$	E_3	A_3
$\vdots$	$\vdots$	$\vdots$	$\vdots$
m	$e_{m1}, e_{m2}, e_{m3}, \ldots, e_{mn}$	E_m	A_m

Step 9: Steps 7 and 8 are repeated and the set of acceptable solutions are obtained. Table 5.3 shows the numerical values for deviation of expert's opinion and accuracy for each acceptable solution. In Table 5.3, e_{ij} ($i = 1$ to n, $j = 1$ to m) is the numerical value assigned to each μ_i for deviation of expert's opinion and E_t ($t = 1$ to m) is the overall quality value calculated for a solution based on deviation of expert's opinion. A_t ($t = 1$ to m) is the numerical value assigned for the level of accuracy attained by each solution.

From the set of acceptable solutions, the solution that satisfies both the criteria with highest possible solution quality is selected as the optimal solution. In some cases, there may be more than one optimal solution leading to a Pareto optimal solution. In a set of Pareto optimal solutions, no solution dominates another solution. In other words, there is no solution in the set which is better (worse) than any other solution from the viewpoint of all the objectives [7].

Step 10: If a satisfactory solution cannot be obtained by the above procedure, there may be a need to modify the operator f.

The proposed method provides a systematic procedure for evaluating all possible solutions and selecting the appropriate one in an interactive manner.

5.5 Application of the Proposed Methodology to Burr Height Estimation

Hazarika et al. [11] applied the proposed methodology in the estimation of burr height in drilling. Burrs are produced during drilling on both entry and exit surfaces of the workpiece due to plastic deformation of the workpiece material. In the present work, heights of entry burrs are measured. The burr height largely depends on the ductility of the material. Figure 5.6 shows the exaggerated view of entry and exit burrs formed during drilling operation. Formation of burrs during drilling is a critical problem which affects surface quality, dimensional accuracy and safety of handling the product. Burrs are unwanted elements and burr removal involves extra cost. Therefore, significant amount of research has been devoted towards prediction and control of burr formation in drilling.

It is evident from the literature that various parameters affecting burr formation in drilling are material properties, process parameters and drill geometry. Drilling burrs can have different shapes and sizes depending on these parameters. Burr shapes and sizes are observed and classified for different workpiece materials based on experimental studies. Min et al. [19] developed a drilling burr prediction and control chart from experimental data. The authors are of the view that drill geometry, process parameters, and material properties affect burr size and shape. Effect of process parameters (feed rate and cutting speed) on burr formation in drilling is widely studied [1, 15, 24]. Feed rate is found to be a significant factor for burr formation in these studies. Burr size is highly affected by the ductility of the workpiece material. More ductile the workpiece material, larger is the burr size. A number of materials with varying ductility are used for experimental studies on burr formation. Pena et al. [20] presented an experimental study on monitoring of burr formation in drilling aluminium workpiece. Effect of cutting speed, spindle speed, feed rate and ductility on burr size is observed. It is observed that burr height increases with increasing ductility. Drill geometry has a significant effect on burr shape and size [13, 14]. Optimization of drill geometry can minimize burr size. The main geometrical parameters of a drill that influence burr size are point angle, chisel edge and corner radius of the cutting edge. In a later work, a methodology is proposed

Fig. 5.6 An exaggerated view of burr formation in drilling

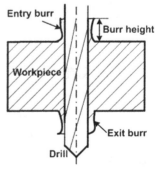

to minimize burr size in drilling by using step drills instead of conventional drills [14]. Burrs formed by a step drill were smaller in size compared to those produced by a conventional drill. For conventional drills, low cutting speeds do not influence burr size. Effect of cutting speed is not very prominent for burr formation compared to feed rate and ductility of the work material. In the present work, burr height is considered as a function of ductility, feed rate and drill geometry. Effect of cutting speed on burr formation is not considered in this study.

5.5.1 Experimental Work

A radial drilling machine (Batliboi Limited, BR618 model) was used for drilling holes in the workpiece in the present work. Three different materials of varying ductility, viz. aluminium, mild steel and cast iron are used as workpiece material. Workpiece is a circular block of diameter 25 mm and height 30 mm. A two flute high-speed steel drill with 10 mm diameter (118° point angle and 30° helix angle) has been used for drilling blind holes of depth 15 mm at different feed rates. For each drilling operation, three replicate experiments were performed in the range of feed rate 104–288 mm/min. Spindle speed and cutting velocity are 800 rpm and 25 m/min respectively. The burr height was measured with an Optical Microscope (Axiotech^{vario} 100 HD, make: Carl Zeiss) of magnification range 5×–200× and supported with KS-300 software. Figure 5.7 shows the photographs of the different equipment used for the experimental work. Tables 5.4, 5.5 and 5.6 show the maximum burr heights for aluminium, mild steel and cast iron work pieces for four different feed rates respectively.

Figure 5.8 shows the maximum value of the burr heights in drilling aluminium, mild steel and cast iron work-pieces with different feed rates. It is evident from Fig. 5.8 that at the same cutting condition, the burr height is the maximum for aluminium which is a ductile material. For mild steel, burr height is lower than

Radial drilling machine   Optical microscope

Fig. 5.7 Machine and equipment used for the experiments

Table 5.4 Burr heights for different feed rates in drilling aluminium workpiece [11]

Feed rate (mm/min)	Maximum burr height (mm)		
	Replicate 1	Replicate 2	Replicate 3
104	0.20	0.18	0.16
144	0.24	0.24	0.23
200	0.36	0.34	0.33
288	0.40	0.38	0.37

Table 5.5 Burr heights for different feed rates in drilling mild steel workpiece [11]

Feed rate (mm/min)	Maximum burr height (mm)		
	Replicate 1	Replicate 2	Replicate 3
104	0.12	0.16	0.12
144	0.21	0.20	0.21
200	0.33	0.29	0.32
288	0.37	0.32	0.35

Table 5.6 Burr heights for different feed rates in drilling cast iron workpiece [11]

Feed rate (mm/min)	Maximum burr height (mm)		
	Replicate 1	Replicate 2	Replicate 3
104	0.05	0.09	0.09
144	0.10	0.10	0.11
200	0.12	0.09	0.09
288	0.13	0.14	0.12

Fig. 5.8 Burr height with different feed rate

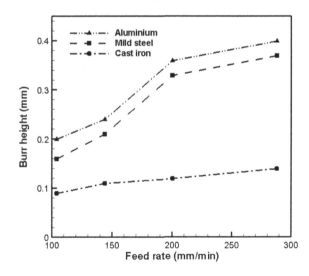

aluminium as mild steel is less ductile than aluminium. For both aluminium and mild steel, there is significant increase in burr height with increase in feed rate. Burr height is very small for cast iron which is a brittle material. Variation of burr height with feed rate is not significant in case of cast iron.

Fig. 5.9 Workpiece and burr
height

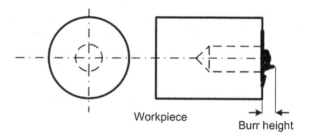

From the experimental study, the maximum and minimum values of burr
heights are found as 0.40 mm (aluminium workpiece at 288 mm/min feed rate)
and 0.05 mm (cast iron workpiece at 104 mm/min feed rate). Moreover, it is
observed that for the replicate experiments, the burr height is varying to some
extent. This is due to the inherent statistical variation in the machining process. In
case of aluminium and cast iron, the maximum variation of burr heights for replicate experiments is found as 0.04 mm. The maximum variation of burr height for
replicate experiments for mild steel is 0.05 mm. Figure 5.9 shows the schematic
diagram of the workpiece and burr height.

5.5.2 Application of the Fine Tuning Methodology

In this section, the fine tuning methodology is applied in estimation of entry burr
height in drilling. To represent different membership grades for burr heights (data
obtained from the experiments), the standard S-function is selected. Figure 5.10

Fig. 5.10 Membership
function for observed burr
heights [11]

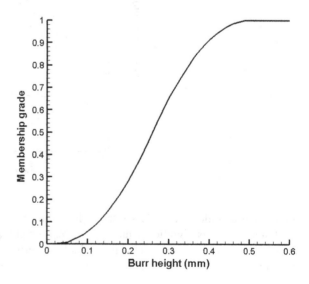

shows the overall membership grades μ_o for the observed burr heights. The value of μ_o for the maximum burr height 0.40 mm is 0.913 and that for minimum burr height 0.05 mm is 0.008. For aluminium, the maximum variation of burr height for replicate experiments is 0.04 mm (at 104 mm/min feed rate) for which the variation in the value of μ_o is 0.11. Thus, there may be an error of the order of 0.11 in the estimation of μ_o for aluminium. For mild steel and cast iron, the maximum variations of burr height for replicate experiments are 0.05 mm (at 288 mm/min feed rate) and 0.04 mm (at 104 mm/min feed rate) and errors in the value of μ_o for mild steel and cast iron may be 0.13 and 0.04 respectively.

With the knowledge acquired from the literature, it is observed that ductility of the workpiece material, feed rate and tool geometry are the three significant parameters that affect burr formation in drilling. As discussed in Chap. 4, the varying effect of these input parameters on burr height is considered by adopting the relation given by Eq. (4.1). The ductility of the workpiece material plays a dominant role compared to the other two parameters. The overall computed membership grade μ_c for burr height is calculated from Eq. (4.1). The initial values of the μ_i (μ_{duc}, $\mu_{feedrate}$ and μ_{tool}) and their variable bounds are provided by the expert. For three different materials of varying ductility and three different feed rates, the values are given in Table 5.7. In the experimental work, a conventional drill with 118° point angle is used for drilling operation. For conventional drills, burr height is found more [14]. Therefore tool geometry (μ_{tool}) is assigned the membership grade 0.9 in Table 5.7.

The overall membership grade μ_c for burr height is calculated from Eq. (4.1) for all the combinations of workpiece material, feed rate and tool geometry. The objective function given by Eq. (5.10) is minimized using optimization technique FMINCON in MATLAB (Version 7). FMINCON attempts to find a constrained minimum of a scalar function of several variables starting at an initial estimate. It uses sequential quadratic programming for optimization. The design variables are the membership grades of individual attributes, i.e. μ_{duc1}, μ_{duc2}, μ_{duc3}, $\mu_{feedrate1}$, $\mu_{feedrate2}$, $\mu_{feedrate3}$ and μ_{tool}. Following the methodology described in Sect. 5.4, each solution is evaluated for accuracy and deviation of expert's opinion. Table 5.8 shows the acceptable solutions satisfying the criteria that the minimum level for accuracy as well as deviation of expert's opinion should be satisfactory. Between Solutions 1 and 2, Solution 1 is better. Among the Solutions 3–7, Solution 3 is the best as it dominates the other solutions. However, between Solutions 1 and 3, no solution dominates the other. Both the solutions form a set of Pareto optimal solution from the viewpoint of satisfying the criteria for accuracy and deviation

Table 5.7 Input parameter membership grades and variable bounds given by the expert

Workpiece material	μ_{duc}		Variable bound	Feed rate (mm/min)	$\mu_{feedrate}$		Variable bound	μ_{tool}	Variable bound
Aluminium	μ_{duc1}	0.9	0.75–0.95	288	$\mu_{feedrate1}$	0.8	0.75–0.90	0.9	0.8–1
Mild steel	μ_{duc2}	0.7	0.60–0.75	200	$\mu_{feedrate2}$	0.5	0.45–0.60		
Cast iron	μ_{duc3}	0.2	0.15–0.25	104	$\mu_{feedrate3}$	0.3	0.15–0.30		

Table 5.8 Acceptable solutions based on accuracy and deviation of expert's opinion [11]

Solution	Overall quality value for a solution for deviation of expert's opinion (E_t)	Numerical value assigned for accuracy (A_t)
1	**9.43**	7
2	8.57	7
3	**8**	**8**
4	7.86	8
5	7.71	8
6	7.43	8
7	7.14	8

Pareto-optimal solutions indicated by the boldfaced

of expert's opinion. These are indicated by boldfaced in Table 5.8. A higher level of decision is required to choose between these two solutions. Table 5.9 shows the values of design variables for Solution 1 and 3.

For validation of the proposed method, drilling experiments were performed at an intermediate feed rate of 144 mm/min. The maximum burr heights of three replicate experiments for aluminium at feed rate 144 mm/min were found as 0.24, 0.24 and 0.23 mm. Corresponding overall membership grade μ_o of these observed burr heights are 0.42, 0.42 and 0.38. The predicted membership grade μ_c is 0.54 with the initial expert's values of μ_i ($i = 1$ to n) which gives a difference of 0.12, 0.12 and 0.16 with the observed μ_o values of the three replicate experiments respectively. However, the value of μ_c is 0.43 with the fine tuned values of μ_i (Solution-3 in Table 5.9) giving a difference of 0.01, 0.01 and 0.05 with the observed μ_o values of the replicate experiments. Thus, there is a better matching of μ_c and μ_o values with fine tuned values of μ_i than with initial expert's values of μ_i. For mild steel, the burr heights of three replicate experiments were found as 0.21, 0.20 and 0.21 mm with corresponding values of μ_o as 0.31, 0.28 and 0.31. The value of μ_c (0.42) with the initial expert's values of μ_i gives a difference of 0.11, 0.14 and 0.11 with observed μ_o values whereas the value of μ_c (0.34) with fine tuned values of μ_i gives a difference of 0.03, 0.06 and 0.03. For cast iron, the burr heights of three replicate experiments were found as 0.10, 0.10 and 0.11 mm with corresponding values of μ_o as 0.06, 0.06 and 0.07. The value of μ_c (0.12) with the initial expert's values of μ_i gives a difference of 0.06, 0.06 and 0.05 with observed μ_o values whereas the value of μ_c (0.04) with fine tuned values of μ_i gives a difference of 0.02, 0.02 and 0.03.

Table 5.9 The optimal solutions for the membership grades of input parameters [11]

Input parameter μ_i	Solution-1	Solution-3
μ_{duc1}	0.95	0.99
μ_{duc2}	0.80	0.80
μ_{duc3}	0.10	0.10
$\mu_{feedrate1}$	0.90	0.95
$\mu_{feedrate2}$	0.60	0.70
$\mu_{feedrate3}$	0.15	0.05
μ_{tool}	0.86	0.81

Thus it is observed that in all the three cases, the fine tuned values of μ_i ($i = 1$ to n) give better matching of μ_c and μ_o than with the initial expert's values of μ_i. Fine tuning of the initial expert's estimates has enhanced the performance of the burr height prediction methodology. The methodology is suitable where limited information is available initially and information value keeps on increasing. The proposed methodology provides a systematic procedure for evaluating all possible solutions and selecting the appropriate one in an interactive manner.

5.6 Conclusion

In this chapter, some methods for the construction of membership grades are described. In addition to methods described, a number of psychological methods can be employed for constructing the membership grades. For example, response time of expert to answer a question regarding the belongingness of some element to fuzzy set may form the one basis of estimating membership grades. A method to fine tune the membership grades assigned by an expert is also described. The method is demonstrated for the drilling burr estimation problem. There is a lot of scope for developing other methods for constructing/fine-tuning of membership grades.

Acknowledgments Significant part of this chapter has been taken from M. Hazarika et al. [11]. We are thankful to NOVA Publishers for allowing us to use this material as per copyright agreement.

References

1. Aramendi G, Arana R, Argelich C (2008) Monitoring of drilling for burr detection using machine learning techniques. Int J Digit Manuf 1:25–32
2. Arslan A, Kaya M (2001) Determination of fuzzy logic membership functions using genetic algorithms. Fuzzy Sets Syst 118:297–306
3. Bagis A (2003) Determining fuzzy membership functions with tabu search-an application to control. Fuzzy Sets Syst 139:209–225
4. Bai SM, Chen SM (2008) Automatically constructing grade membership functions of fuzzy rules for students' evaluation. Expert Syst Appl 35:1408–1414
5. Chen MS, Wang SW (1999) Fuzzy clustering analysis for optimizing fuzzy membership functions. Fuzzy Sets Syst 103:239–254
6. Choi BI, Rhee FCH (2009) Interval type-2 fuzzy membership function generation methods for pattern recognition. Inf Sci 179:2102–2122
7. Dixit PM, Dixit US (2008) Modeling of metal forming and machining processes. Springer, London
8. Furukawa M, Yamakawa T (1995) The design algorithms of membership functions for a fuzzy neuron. Fuzzy Sets Syst 71:329–343
9. Garibaldi JM, Ifeachor EC (1999) Application of simulated annealing fuzzy model tuning to umbilical cord acid-base interpretation. IEEE Trans Fuzzy Syst 7:72–84
10. Hazarika M, Dixit US, Deb S (2010) A method for fine tuning the membership grades assigned by experts: an application to burr height estimation in drilling. In: Proceedings of the 3rd international and 24th AIMTDR conference, A.U. College of Engineering, Visakhapatnam

11. Hazarika M, Dixit US, Deb S (2012) A method for fine tuning the membership grades assigned by experts: an application to burr height estimation in drilling. J Manuf Technol Res 4(1,2):75–87
12. Hong TP, Lee CY (1996) Induction of fuzzy rules and membership functions from training examples. Fuzzy Sets Syst 84:33–47
13. Ko SL, Lee JK (2001) Analysis of burr formation in drilling with a new-concept drill. J Mater Process Technol 113:392–398
14. Ko SL, Chang JE, Yang GE (2003) Burr minimizing schemes in drilling. J Mater Process Technol 140:237–242
15. Lauderbaugh LK (2009) Analysis of the effects of process parameters on exit burrs in drilling using a combined simulation and experimental approach. J Mater Process Technol 209:1909–1919
16. Liu X, Pedrycz W (2007) The development of fuzzy decision trees in the framework of axiomatic fuzzy set logic. Appl Soft Comput 7:325–342
17. Medaglia AL, Fang SC, Nuttle HLW, Wilson JR (2002) An efficient and flexible mechanism for constructing membership functions. Eur J Oper Res 139:84–95
18. Medasani S, Kim J, Krishnapuram R (1998) An overview of membership function generation techniques for pattern recognition. Int J Approximate Reasoning 19:391–417
19. Min S, Kim J, Dornfeld DA (2001) Development of drilling burr control chart for low alloy steel, AISI 4118. J Mater Process Technol 113:4–9
20. Pena B, Aramendi G, Rivero A, Lacalle LNL (2005) Monitoring of drilling for burr detection using spindle torque. Int J Mach Tools Manuf 45:1614–1621
21. Saaty TL (1974) Measuring the fuzziness of sets. J Cybern 4:53–61
22. Saaty TL (1977) A scaling method for priorities in hierarchical structures. J Math Psychol 15:234–281
23. Sancho-Royo and Verdegay JL (1999) Methods for the construction of membership functions. Int J Intell Syst 14:1213–1230
24. Stein JM, Dornfeld DA (1997) Burr formation in drilling miniature holes. Ann CIRP 46:63–66
25. Yang CC, Bose NK (2006) Generating fuzzy membership functions with self-organizing feature map. Pattern Recogn Lett 27:356–365
26. Zadeh LA (1976) A fuzzy-algorithmic approach to the definition of complex or imprecise concepts. Int J Man Mach Stud 8:249–291

Chapter 6
Fixturing Consideration in Setup Planning

Abstract It is better to obtain a conceptual design of jigs and fixtures at setup planning stage itself. A brief review of different types of fixtures and relevant research is presented. Fixture design requires information about machining force. These can be estimated in an approximate manner using physics based or soft computing based approaches. Design of locators and clamps is based on the principles of contact mechanics. Considering uncertainty, fuzzy arithmetic can be used for the estimation of forces and design dimensions. The location of clamps and locators needs to be optimized for gaining better stiffness of the system. An illustrative example on end milling demonstrates how fixture related information can be obtained along with setup planning.

Keywords Jigs · Fixtures · Machining forces · Clamps and locators · End milling · Contact mechanics

6.1 Introduction

Setup planning is a critical part of process planning for machining a component and it is discussed in detail in the previous chapters. Generally, setup planning systems provide the optimum number of setups to machine a component, machining operation sequences, setup sequences, and datum for each setup. However, the output of the traditional setup planning approaches is limited and insufficient for upstream process planning activity such as fixture design. In fact, fixturing requirements are to be considered as an integral part of setup planning. To generate a robust and practical solution for machining a component, setup, fixturing and machining constraints are to be considered simultaneously. It is essential to estimate the machining forces, clamping forces and the range of process parameters during setup planning stage considering the feasibility of fixturing. In absence of this, the independent fixture design module may fail to generate feasible fixture

© The Author(s) 2015
M. Hazarika and U.S. Dixit, *Setup Planning for Machining*,
SpringerBriefs in Manufacturing and Surface Engineering,
DOI 10.1007/978-3-319-13320-1_6

plans leading to the need of redesigning of setups. Proper integration of setup planning and fixturing can give practical and complete solutions to setup planning. A discussion on different fixtures used for machining is presented in the following section.

6.2 Different Types of Fixtures and Brief Review on Fixture Design

Fixtures are used in manufacturing for locating, supporting and securing the work piece in the correct position with respect to the machine tool during machining. Three important functions a fixture has to perform are part location, support and immobilization or restraining of the part to be machined. A fixture may be a single device or a combination of components. Generally, all fixtures consist of locators, clamps and supports. A locator is usually a fixed component of a fixture that restrains the degrees of freedom of movement of a part. A clamp exerts force that holds a part securely in the fixture to resist all other external forces. Clamps are either manually operated or actuated by pneumatic or hydraulic power. A support is a fixed or adjustable element of a fixture placed under the part to be machined in cases when operational forces and possibilities of part deflection are more. Locating and clamping mechanisms provide support and maintain the work piece in a particular position in a setup and resist gravity and other operational forces. The purpose of setup and fixturing is to ensure the stability and precision of the workpiece during machining processes. The geometry of a part plays a key role in the selection of the type of fixtures to be used for machining. Selection of fixture also depends on the product variety and volume of production. Two main types of fixtures mostly used in the industry are dedicated/custom made fixture and modular fixture. Fixtures used for prismatic and rotational parts are of different types.

Variety of fixtures is used for fixturing prismatic parts. They are dedicated/custom made fixtures, vices, modular fixtures, etc. A vice is the simplest type of fixture which is the most widely used. A vice uses a stationary jaw for locating the part and a movable jaw is pressed against the part to clamp it tightly. Use of vice is common both in vertical and horizontal machines. However, vices are inflexible and restricted by their sizes. Dedicated fixtures can be used only for those components for which they are designed. Dedicated fixtures are costly, inflexible and time consuming in designing. In this era of high product variety and shorter product life, use of dedicated fixtures increases the product cost and lead time. A better alternative to dedicated fixture is modular fixture. Use of modular fixture is inevitable to keep in pace with the demand for flexibility and product variety coupled with increasing design complexity. A modular fixturing system contains a set of standard fixturing elements such as a flat base plate, locators, clamps, V-blocks, and supporting elements. Modular fixture components are assembled on the base plate. Predetermined grids of holes are drilled in the base plate and fixturing elements like studs, clamps, locators, etc. are fitted in these holes to fixture the component to be machined. Unlike dedicated fixtures, they can be re-used and adapted for different parts. A modular fixture can be disassembled into its parts after machining of a component and reassembled again for machining of another component thus adding flexibility. Modular fixtures are capable of handling a wide variety of part with varying sizes and shapes. Moreover, they are cost effective and time efficient. These qualities have made modular fixtures the most preferred choice for machining of prismatic parts.

Different features of a prismatic part are used for locating and clamping. By location, the position and orientation of a part is established relative to the machine tool. The part is held in the required position with clamps against the locators during machining. Clamps exert a clamping force to restrain the part to be machined. Locator supporting components and clamp supporting components are used wherever necessary. The fixture layout plan gives the detail of the part's surfaces to be used for locating and clamping as well as exact positions of the locating and clamping points. The location of a part should be such that the six degrees of freedom are adequately constrained during machining. The six degrees of freedom are translations along x, y and z axes and rotations about these three axes as shown in Fig. 6.1. Generally three types of features are used for location in case of prismatic parts—planar surfaces, holes and external profiles. There are mainly two locating methods for prismatic parts—3-plane locating (3–2–1) and 1-plane and 2-hole locating. For a prismatic part, the planar surfaces of the part can be used conveniently for 3–2–1 locating and this locating method is mostly

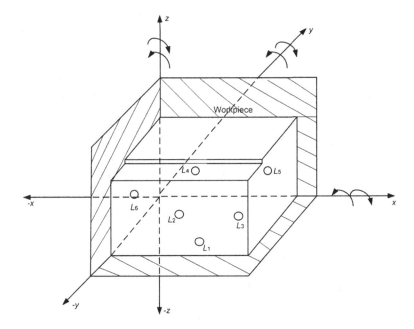

Fig. 6.1 The six degrees of freedom restrained by 3–2–1 location

used. In Fig. 6.1, six locators are used in three datum faces following 3–2–1 locating method to locate the prismatic component. Note that locators L_1, L_2 and L_3 touch the bottom surface, the locators L_4 and L_5 touch the surface normal to y-axis and locator L_6 touches the surface normal to x-axis. Normally, the largest surfaces opposite to the locating surfaces are used for clamping. The position of the clamp should be so chosen that there is no interference between the clamping elements and the cutting tool.

Different fixtures used for rotational parts are jaw chucks, drill chucks, face plates, collets, etc. Mainly chuck-type fixtures are used for rotational parts. In these fixtures, radially adjustable jaws are used to define the axis of rotational part. Generally for a rotational part, end faces are selected as locating features and external cylindrical faces are selected as clamping features. For details of different locating and clamping methods, the reader is directed to Joshi [17].

Jigs are complementary to fixtures and guide the cutting tools. For example, in duplicating a key, original key can be used a jig to guide the tool into a specified path. Some types of jigs are also called templates. Modern computer numerical controlled (CNC) machines may not require jigs as the tool path is pre-programmed.

Fixture design is a very important part of process planning. It consists of a number of steps, viz. fixture planning, fixture layout design, and detailed fixture design, i.e. individual fixture element design. In fixture planning stage, the type of fixture to be used, orientation of the part and possible datum features for locating, clamping and supporting faces are decided. The second phase, the fixture layout design is also called conceptual fixture design phase. It gives specific positions and types of locators, clamps and supports and clamping sequence as well as clamping forces. In the next phase, each of the fixture elements: locators, clamps and supports are designed in detail.

Extensive research work is available in the literature on fixture design. All the aspects of fixture design, viz. fixture-workpiece contact condition, elastic deformation and total restraint of the workpiece, stability analysis, quasi-static fixture-workpiece system as well as dynamic fixture-workpiece system, clamping sequence and clamping force, etc. are adequately addressed by the researchers. However, these works are concerned with fixture design only, without any integration with setup planning. Many researchers consider fixturing requirements as an integral part of setup planning. There are some attempts in the literature to develop setup plans considering fixturing aspect. The pioneering work integrating setup planning and fixturing can be traced back to Boerma and Kals [3, 4]. They developed a system called FIXES for automatic selection of setups and datum based on feature tolerances for machining of prismatic parts. The system automatically selects the positioning, clamping and supporting faces for each setup. There have been continuous efforts since then to integrate setup planning and fixturing [10, 14, 16, 18, 24, 28, 29]. Young and Bell [29] proposed a methodology for integrating technological and geometrical information of the part and fixturing constraints for automating setup planning. The method uses machine capability, part geometry, precedence relations and tolerances among the features as the main constraints for setup planning. Sakurai [24] used algorithmic and heuristic methods for automating setup planning and fixture design for prismatic parts. Setups are formed on the basis of TAD analysis of the features and some heuristics are applied to find the best locating faces for each setup. Some of the criteria considered for setup planning and fixturing are part geometry, precedence constraints, total restraint of the part during machining, interference checking for fixturing, and minimum part deformation. Tseng [28] proposed an approach for fixture design analysis for feature based machining of prismatic parts. The method analyses the setups and related fixturing requirements in a sequential feature-based machining considering operation precedences. The workpiece shape at an intermediate step and the feature to be cut is given as input to the fixturing analysis module. The output includes locating faces and points, clamping points, and feasible height ranges for locating and clamping devices. Joneja and Chang [16] combined setup and fixture planning considering part geometry, precedence constraints and restraint of the machined part. The setup planning module groups the surfaces to be machined into different setups, generates alternative setup plans, and selects the plan with the minimum number of setups. The fixture planning module decides the clamping method (vise or modular fixture) and locating, clamping, and supporting faces were decided after tool interference checking. Kaya and Ozturk [18] presented an algorithmic approach to develop an integrated system to generate machining operation groups for different setups and fixture configuration layout for each setup. Feature precedence matrix is used to form the setup and machining operation sequence. An algorithm is presented for selecting the locating and clamping positions. Time varying dynamic machining force analysis is carried out to ensure workpiece stability against cutting and clamping forces. Finite element analysis (FEA) technique is used for stability analysis of the workpiece. Huang and Xu [14] proposed a methodology for setup planning and datum selection for machining prismatic parts with total integration of the part features, machining operations, tools, and fixture modules. Parallelism and perpendicularity tolerances among the features are considered for feature clustering and setup formation. Objectives incorporated into the setup formation algorithm are minimizing the number of setups and ensuring that the features with tight tolerance relations are machined in the same setup. Gologlu [10] developed a knowledge-based methodology for setup planning and datum selection incorporating machining and fixturing constraints. Part geometry, tolerance and dimensions of the part, and feature interaction are used to form the precedence constraints among the features. Setup clusters were formed based on TAD and precedence relations. Conceptual fixture design is performed and locating surfaces for the setups are identified.

Some recent works on integrated setup planning and fixturing are discussed in this paragraph. With growing need for automation in all fields of manufacturing, use of standard data exchange format for effective communication of product data is increasing. For convenient exchange of product data among different CAD/CAPP/CAM systems, data exchange standards like IGES, STEP are being used for integrated setup planning and fixturing approaches [2, 5]. Bansal et al. [2] suggested a modular fixture planning system integrated with feature recognition module, setup planning module and tolerance analysis module. They designed an interactive user interface to take direct input from the STEP file about the component geometry for feature recognition. After reconstruction of the part, setups are formed based on TAD, feature tolerance relations, and machine and tool spatial constraints. Acceptable locating points for each setup are searched subject to accessibility, stability, and minimum tolerance criteria. Borgia et al. [5] used a STEP-NC compliant data structure and proposed a method for setup planning and fixturing based on mathematical programming. The authors used tombstone pallets as fixture and work piece is clamped to the pallet surface for machining. Emphasis is given on maximum use of the pallet surface and mininum number of setups. Considering the state of the art, a four-axis CNC machining center with rotary table is used. There have been attempts for integrated setup planning and fixturing for machining of box-shaped parts. Stampfer [26] developed an automated setup and fixture planning method for box-shaped parts. Features to be machined in a setup, setup sequences, and locating and clamping surfaces for each setup can be generated automatically by the proposed system. Setup formation is done based on the tolerance relations among the features. An orientation of the workpiece is searched where maximum number of features can be machined in a particular clamping position. Locating and clamping surfaces are selected based on shape, size, and position of the surfaces. A similar approach for machining box-shaped parts is found in Attila et al. [1] that takes the CAD model of the part to be machined along with its related data as input. The output gives the total number of setups in their proper sequence along with the appropriate fixtures in the form of a CAD model. The proposed system consists of four modules: the CAD model post-processing/fixture pre-processing module, the setup and fixture planning module, the operation-planning module and the fixture configuration module. The output of the preceding module becomes the input to the succeeding module. The CAD model of the part to be machined is stored in IGES format. The first module analyses the CAD model and extracts the different features of the part with their related information, e.g. tolerances, dimensions, shape, precision, etc. Moreover it attempts to suggest some conceptual fixturing solutions. The setup and fixture planning module analyses these conceptual fixturing solutions and further modifies it based on achieving prescribed tolerances and minimum number of setups. All locating and clamping surfaces are critically examined in this module. The fixture configuration module attempts to provide a fixturing solution by selecting the appropriate supporting, locating and clamping elements and properly positioning them relative to the part. The suggested CAD models of the fixtures can be viewed in Solid Edge environment and necessary modifications can be made if there is a change in design. Recently, more emphasis is given on type and design of machine tool for efficient and interference free fixturing and setup planning [6, 13, 19]. Cai et al. [6] developed a methodology for setup planning of prismatic parts which is adaptable to different multi-axis machines. A kinematic model for tool accessibility is proposed to generate the feasible tool movement space for different multi-axis machines. Initially, setups are formed based on TAD for a three-axis machine which becomes the input for other machines of four-axis and five-axis type. Number of setups for the same part can be reduced with four-axis, five-axis machines, as some setups can be merged with more options for tool accessibility in the multi-axis machines. Hu et al. [13] focussed on the problem of obstacle-free accessibility for machining the features with minimum number of setups in case of five-axis NC machining. Obstacles like fixtures and clamps are also considered in addition to static obstacles like part surface for determining the tool orientation and accessibility. A heuristic-based solution for finding a minimum number of set-ups for machining the part surface without colliding with the obstacles is obtained. The proposed approach also assures correct tool orientations avoiding fixtures at the stage of tool path planning. Leonesio et al. [19] proposed setup planning and fixture selection by integrating CAD model analysis of the workpiece and machine tool design. CAD model analysis involves identification of machining features and the selection of the best suited machining operations to machine those features. Machine tool design module gives information on general-purpose machine tools that meet the machining requirements considering machine tool dynamic and kinematic constraints. Combining both, dynamic cutting simulation is performed and the dynamic behaviour of machine tools is evaluated considering energy consumption, tool wear, surface roughness and required spindle power and torque. Based on these previous steps, selection of fixtures, workpiece orientations and setups can be decided.

There are limited attempts on developing a setup planning strategy that provides sufficient input to the fixture design and further process planning. Most of the works in the literature considering fixturing deal with the conceptual fixture design phase by identifying the datum features. The aspects relating to position of the locators and clamps, machining force, clamping force, and range of process parameters require more attention.

In view of it, in this chapter, an example of integration of setup planning with fixturing is presented. Fixturing requirements can be incorporated into the developed setup planning expert system. In addition to the setup planning information, the system now can provide the following output: recommended depth of cut/feed in fuzzy form, machining and clamping forces in fuzzy form, approximate optimal locator and clamp layout and sizes of the locators and clamps. The fixture designer can further optimize the fixture plan by taking input from the setup planning module. This methodology helps in improving the overall efficiency of the process plan. Moreover, the uncertainties associated with the work material, clamp material and clamping torque are considered by means of fuzzy arithmetic.

6.3   The Architecture of the Fixturing Information Generation Module

A block diagram representation of the setup planning system with the detail of fixturing information generation module is shown in Fig. 6.2. The fixturing information generation module contains the sub-modules B, C, D and E and it takes input from module A (described in Chap. 4) that generates number of setups, operation sequences, and datum for each setup. B is machining force calculation module, module C generates optimized locator and clamp layout, D is the workpiece-fixture contact module, and module E selects locator and clamp sizes.

6.3.1 Setups, Operation Sequence and Datum Selection Module A

In module A of Fig. 6.2, setup generation, operation sequencing and datum selection are carried out. A typical module was described in detail in Chap. 4. This module may be based on expert system and may use fuzzy set theory. The module may have generative or variant systems. It may also include both the systems and make use the best of both the systems.

6.3.2 Approximate Machining Force Calculation Module B

Module B calculates approximate machining forces. There are various approaches to calculating the force; however, all of them provide only a rough estimate of machining forces. This is due to limited knowledge of the physics of machining process and difficulty in getting accurate input parameters of the machining model. Fortunately, at the fixture design stage, approximate values of machining forces are sufficient, because designer employs the factor of safety as in done in the design of any engineering product.

One approach for estimating the machining forces is total specific energy approach [25, 27]. The energy consumption per unit volume of material removal

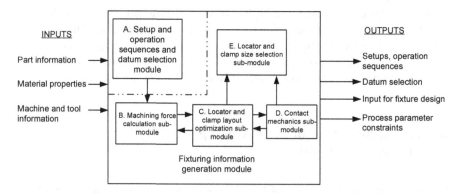

Fig. 6.2 The setup planning system with the detailed fixturing information generation module. With permission from Hazarika et al. [11]. Copyright [2010] Springer

is called specific energy. In general, the specific energy is dependent on the uncut chip thickness. The smaller is the chip thickness, the greater is the specific energy. The data for specific energy for a particular material is available in many machining data handbooks. With the specific energy approach, the main cutting force can be easily calculated. The thrust force can be taken as 0.3–0.5 times the main cutting force. The thrust force can be resolved into forces normal to machined surface and along feed direction.

A number of mechanistic models are available in literature. In mechanistic modelling, the complex cutting tool is divided on slices along its axis and the cutting forces are locally modelled by simple analytical formulae. Finally, these local cutting forces are integrated to provide overall values of machining forces. The drawback of this approach is unavailability of input information.

Several researchers have estimated machining forces by finite element method (FEM). There are three ways of FEM modelling—Eulerian, Lagrangian and Arbitrary Lagrangian Eulerian (ALE) methods. In the Eulerian method, the mesh is spatially fixed while the material is allowed to flow through meshed control volume. The advantage of the Eulerian method is that the element distortion is absent as the mesh is fixed. The drawback is that the initial shape of the chip and contact conditions should be known a priori. On the other hand, in the Lagrangian method, the mesh is attached to the workpiece and the elements are allowed to deform similar to actual machining. This method required a well-defined chip separation criterion. The excessive mesh distortion, need for frequent re-meshing and a large amount of computational times are the drawback of this method. In ALE method, the mesh is neither spatially fixed nor attached to the material. Instead it is allowed to flow with the material. In this manner, severe distortion of elements is avoided without the need for re-meshing. However, FEM is also not a very convenient choice for machining force estimation. Firstly, it takes a lot of computational time. Secondly, the results are not very accurate due to improper information abut physics of machining, material behaviour and friction.

The machining forces can also be estimated by using empirical relations, expert system and soft computing based methods. A number of papers have been published on the fuzzy set and neural network based approaches for machining force estimation [8]. These methods require a sufficient amount of shopfloor data.

6.3.3   Locator and Clamp Layout Optimization Module C

Objective of a good fixture configuration design is to design and place the locators and clamps on the workpiece faces at such positions that the passive reaction forces are kept to a minimum. Module C finds the optimized locator and clamp layout which gives the smallest passive locator reaction forces maintaining the workpiece-fixture system stability and minimum deformation condition. Standard 3–2–1 locating principle for machining prismatic parts may be followed. The objective function is formulated to minimize the maximum L_2 norm of locator reaction forces during machining and clamping. Clamping and machining forces along with the part weight are the active known inputs and the locator reactions are the variables to be determined. The following constraints are used in the optimization problem:

Static equilibrium constraint: The necessary and sufficient condition to ensure static equilibrium of the workpiece is to satisfy the force and moment equilibrium equations where the forces and moments consist of the machining forces, clamping forces, part weight and locator reaction forces in the normal direction.

Workpiece-fixture contact constraint: The static equilibrium constraint keeps the workpiece stable during machining. However, it does not account for workpiece slippage or detachment from the locators resulting in negative or zero locator reaction force. Locators must be maintained in contact with the workpiece throughout the machining process to ensure complete immovability of the workpiece. A constraint that all locator reaction forces must be positive takes care of immovability of workpiece.

A frictionless contact between the workpiece and fixture elements may be assumed. A frictionless analysis leads to a conservative and safer fixture design. Coulomb friction generates additional restraint to a workpiece-fixture system. Liao and Hu [22] confirmed with a comparative analysis that a frictionless model predicts higher value of required clamping force than the model considering frictional effects.

The complete locator and clamp layout optimization model can be expressed mathematically as

$$\text{Minimize}\left[\text{maximum of } \sum_{i=1}^{6} R_i^2 \text{ over the entire cutting path}\right], \qquad (6.1)$$

subject to the static equilibrium constraint

$$\sum F = 0, \quad \sum M = 0 \qquad (6.2)$$

and workpiece-fixture contact constraint

$$R_i > 0 \qquad (6.3)$$

where R_i ($i = 1$–6) is the locator reaction force in the normal direction, and $\sum F$ and $\sum M$ are net forces and moments due to machining forces, clamping forces, part weight and locator reaction forces in the normal direction. Modular fixture elements can be used. In that case, an additional constraint that the locating and clamping point coordinates take only discrete values is to be incorporated in the fixture layout optimization model. An integer programming approach can be adopted.

This module can also make use of FEM. FEM can predict detailed information, not only for the passive forces but also deformations. It also can take care for the presence of redundant locators for providing extra-stiffness to the entire system.

6.3.4 Workpiece-Fixture Contact Module D

Contact mechanics approach is used for modelling workpiece-fixture contact conditions in module D. Hertz's contact model can be used to represent the most common cases of contact between the workpiece and the fixture. It gives the contact area, contact deformation and total compressive load for two elastic bodies in contact. The following key assumptions are made so that contact mechanics [15] approach can be used for modelling:

- The workpiece and fixture elements are linear elastic bodies.
- Workpiece-fixture contact area is circular and radius of the contact area is much smaller compared to the radii of curvature of the two contacting bodies.
- The contact deformation is small and is independent of the contact pressure at other contact points.

Spherical locator and clamp contact surfaces are used in this work. The stiffness of the locators and clamps are assumed to be higher than the workpiece stiffness. Hertz's contact model is used to represent the elastic contact between spherical locators/clamps and planer workpiece surface. Some contact mechanics based solution approaches for optimal clamping and contact forces for minimum deformation are found in Li et al. [21], Li and Melkote [20] and Deng and Melkote [9], albeit these papers are concerned with fixture design only, without any integration with setup planning.

In the present module, the normal contact deformation δ_n due to normal force P acting between a spherical-tipped fixture element and planer workpiece surface is obtained from Hertz's contact model [15] as

$$\delta_n = \left[\frac{9P^2}{16RE^2} \right]^{1/3} \tag{6.4}$$

and the normal load P_y to initiate yield in the workpiece material is given by the expression [15]

$$P_y = \frac{\pi^3 R^2}{6E^2} (1.6Y)^3 \tag{6.5}$$

where $1.6Y$ is the maximum contact pressure at workpiece-fixture interface according to von Mises' as well as Tresca' yield criteria. Here Y is the yield stress of the workpiece material in compression.

In Eqs. (6.4) and (6.5)

$$\frac{1}{R} = \frac{1}{R_w} + \frac{1}{R_f} \tag{6.6}$$

$$\frac{1}{E} = \frac{1 - v_w^2}{E_w} + \frac{1 - v_f^2}{E_f} \tag{6.7}$$

where R_w and R_f are the radii of curvature of the workpiece and fixture element contact surface, R is the equivalent radius of curvature of the two bodies in contact. E_w and E_f are the Young's moduli of elasticity of the workpiece and fixture element and E is the equivalent Young's modulus of elasticity; and v_w and v_f are the Poisson's ratios of the workpiece and fixture element respectively.

6.3.5 Locator and Clamp Design Module E

Module E calculates the proper size of the clamps and the locators. Size of a clamp depends on the magnitude of the machining force it has to experience and the tensile strength of the clamp material [7]. For a screw clamp to apply a clamping force F_{clamp}, the clamp nominal diameter d_{clamp} can be found from the expression [12]:

$$d_{clamp} = \frac{T}{0.2\,F_{clamp}} \tag{6.8}$$

where T is the torque applied at the head of the clamp screw. The minimum value of diameter of a screw clamp is found from the following relation [23]:

$$\frac{T}{J} = \frac{2\tau}{d_{clamp}} \tag{6.9}$$

where J is the polar moment of inertia of the clamp screw, τ is the allowable shear stress of the clamp material. Diameter d_{clamp} calculated from Eq. (6.8) should be greater than d_{clamp} calculated from Eq. (6.9).

Figure 6.3 shows the different parameters of a spherical locator. The relations among the different parameters of a spherical locator are given as [12]:

Fig. 6.3 The parameters of a spherical locator. With permission from Hazarika et al. [11]. Copyright [2010] Springer

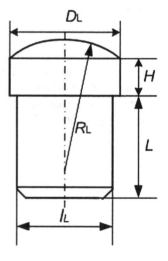

$$H = \frac{1}{3}D_{\mathrm{L}} \text{ to } D_{\mathrm{L}} \tag{6.10}$$

$$R_{\mathrm{L}} = \frac{3}{2}D_{\mathrm{L}} \tag{6.11}$$

$$l_{\mathrm{L}} = L = \frac{3}{4}D_{\mathrm{L}} \tag{6.12}$$

6.4 The General Methodology for Generating Fixturing and Process Related Information

The following steps can be executed for generating the fixturing and process related information:

- Given information of the part, machining operation, machines and tools as input to module A, the setup and machining operation sequences and datum for the setups are obtained.
- The machining forces are computed. Considering the uncertainty in the physics of the process and input data, a fuzzy set based approach may be adopted. In fuzzy set based approach, the required input parameters such as specific cutting energy may be taken as fuzzy numbers. Fuzzy arithmetic may be used for computing the forces as fuzzy numbers.
- To begin with, the clamping force may be taken as the maximum value of machining forces.
- Locator and clamp layout optimization module can be used for calculation. If feasible solution is not obtained, the clamping force is gradually increased till a feasible solution is obtained.
- A suitable value of safety factor (about 2) is still needed for clamping force, considering the uncertainty present in the machining process.
- Clamp diameter can be calculated considering clamping torque and clamping force as fuzzy numbers. Alternatively, the factor of safety approach can be adopted.

Hazarika et al. [11] have suggested a strategy for obtaining the radius of curvature of spherical clamp, R_{clamp}, shown in Fig. 6.4. According to it, Eq. (6.5) can be used for finding out the minimum value of R_{clamp}. To find the maximum value, the following strategy is developed. Figure 6.4 shows the clamp parameters. Here r_{clamp} is the clamp radius, s is the height of the spherical clamp tip and R_{clamp} is the radius of curvature of the spherical clamp. From Fig. 6.4,

$$R_{\mathrm{clamp}}^2 = (R_{\mathrm{clamp}} - s)^2 + r_{\mathrm{clamp}}^2 \tag{6.13}$$

Neglecting very small terms, Eq. (6.13) can be written as

$$R_{\mathrm{clamp}} = \frac{r_{\mathrm{clamp}}^2}{2s} \tag{6.14}$$

Fig. 6.4 The parameters
of a spherical clamp. With
permission from Hazarika
et al. [11]. Copyright [2010]
Springer

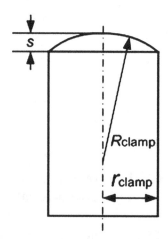

The contacting surface of the workpiece is considered to be a rough surface. With
an objective of proper contact between the spherical clamp and workpiece, s must
be at least equal to the peak to valley roughness height R_t of the workpiece surface. Hence, the maximum value of radius of curvature $(R_{clamp})_{max}$ is given by

$$(R_{clamp})_{max} = \frac{r_{clamp}^2}{2R_t} \qquad (6.15)$$

The deflection δ of the locator on the primary datum under the part weight and
other external forces is given by the relation

$$\delta = \frac{P_L L}{A E_L} \qquad (6.16)$$

where P_L is the total load on the locator, L is the locator height, A is the cross-sec-
tional area of the locator and E_L is the Young's modulus of elasticity of the locator
material. Using Eq. (6.16), strength formulae and typical geometry standards (like
Eqs. 6.10–6.12), one can find out the dimensions of locator.

6.5 An Example of End Milling Process

An end milling operation is used to machine the top face PQRS of the workpiece
shown in Fig. 6.5. From module A, primary datum for the setup is selected based
on tolerance relation, surface area and surface quality. The largest face perpendicular to the primary datum is the secondary datum. The tertiary datum is perpendicular to both the primary and secondary datum. The workpiece is a prismatic block
of dimensions $70 \times 60 \times 50$ mm^3 and the workpiece material is AISI 1018 steel.
Taking the density of AISI 1018 steel as 7.87 g/cc, the weight of the workpiece is

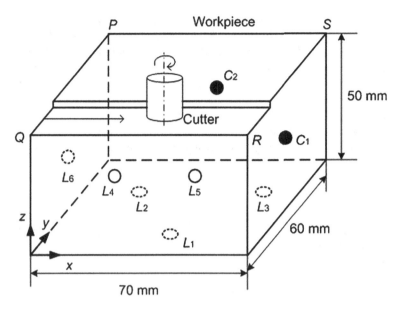

Fig. 6.5 The end milling of the example part. With permission from Hazarika et al. [11]. Copyright [2010] Springer

found to be 16.5 N. It is fixed with three locators L_1, L_2 and L_3 on the primary datum (x-y plane), two locators L_4 and L_5 on the secondary datum (x-z plane) and one locator L_6 on the tertiary datum (y-z plane). Clamp C_1 is placed opposite to the locator on the tertiary datum and C_2 is placed opposite to the locators on the secondary datum. Spherical locator and clamp contact surfaces are used in this work for proper contact with rough workpiece surface. Screw clamps made of 2340 medium carbon alloy steel are selected. Locator material is water hardening steel W1 with 0.6 % carbon content. Workpiece, clamp and locator material properties are taken from standard data books. A 20 mm diameter helical end mill with four flutes and 30° helix angle is used for the milling operation. A torque of 2,000 N-mm is applied at the head of the clamp screw with one hand operation. The central line average (CLA) surface roughness height of the workpiece contact surface is considered as 50 μm (N12).

The material parameters can be taken as fuzzy numbers. For that purpose, low, medium and high estimates are obtained. The basis of these estimates may be that the variations in specific cutting energy and yield stress of the workpiece material may go up to ±30 and ±10 % respectively. The Young's moduli of elasticity of the workpiece, clamp and locator materials may vary by ±5 %. The variations in clamping torque and peak to valley roughness height of the workpiece contact surface are considered as ±10 %. Linear triangular fuzzy membership functions are assumed for these parameters. A linear triangular membership function is constructed by taking the membership grade as 1.0 at most likely (m) and 0.5 at low (l) and high (h) estimates of a parameter. With these three points a triangle is constructed for each parameter.

The machining and clamping forces are obtained as fuzzy numbers. Figure 6.6 shows the fuzzy machining forces F_x, F_y and F_z and clamping force F_{clamp} at different membership grades for the end milling operation at 0.5 mm depth of cut and 0.1 mm/tooth feed. Standard 3–2–1 location with one clamp each on secondary and tertiary datum is followed in this case. From Fig. 6.6a, the high (h) estimates of F_x, F_y and F_z at 0.5 membership grade are 139.89, 99.76 and 28.50 N respectively. From Fig. 6.6b, the most likely (m) value of clamping force at membership grade 1.0 is 445.68 N and the low (l) and high (h) estimates at 0.5 membership grade are 280.88 and 643.50 N respectively. Designing for the worst case condition, high

Fig. 6.6 Membership function for **a** machining forces, **b** clamping forces at 0.5 mm depth of cut and 0.1 mm/tooth feed. With permission from Hazarika et al. [11]. Copyright [2010] Springer

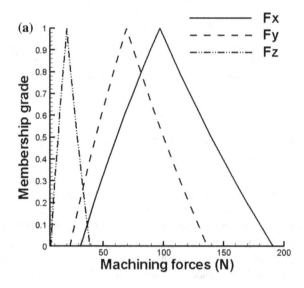

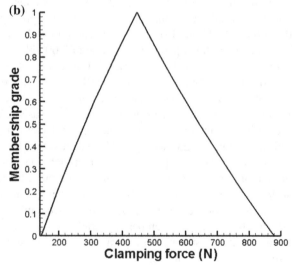

estimate of clamping force 643.50 N at membership grade 0.5 is considered. Radius of the clamp r_{clamp} ($d_{clamp}/2$) is found as 7 mm from Eq. (6.8). The value of r_{clamp} is greater than the minimum value of r_{clamp} (4.82 mm) found from Eq. (6.9). For finding out the radius of curvature R_{clamp} for spherical clamp, minimum and maximum radii of curvature obtained from Eqs. (6.14) and (6.15) respectively are made equal. The value of R_{clamp} comes to be 98 at 0.5 mm depth of cut.

Considering a very small deflection of 0.001 mm of the locator under the part weight and other external forces, locator diameter is found as 12 mm. Spherical locator button diameter D_L is calculated as 16 mm. Radius of curvature of the spherical locator button, R_L comes to be 24 mm from Eq. (6.11). However, in the proposed design, R_L is found considering the onset of yielding in the workpiece material. The minimum value of R_L is calculated as 87 mm. Height of the locator button, H and height of the locator, L are found as 8 and 12 mm respectively from Eqs. (6.10) and (6.12). H is considered half of the button diameter D_L.

Figure 6.7 shows the upper bound of depth of cut at membership grades 0.5 and above for worst case design condition. High estimates of F_x, F_y, F_z and F_{clamp} at 0.5 membership grade are considered. Upper bound for depth of cut at 0.5 membership grade is 0.5 mm and at 1.0 membership grade, it is 0.722 mm.

The approximate relation between depth of cut d, feed f and cutting force F_c can be expressed by the following expression

$$F_c = kfd \qquad (6.17)$$

where k is the proportionality constant. Variable bounds for feed can be calculated using Eq. (6.17) by the proposed method.

Machining and clamping forces increase with higher value of depth of cut and feed. To use higher value of depth of cut or feed, two clamps on one face may be used so that clamping force is reduced on each clamp. Figure 6.8 shows the upper

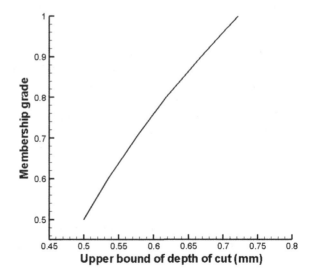

Fig. 6.7 Membership function for upper bound of depth of cut for single clamp design. With permission from Hazarika et al. [11]. Copyright [2010] Springer

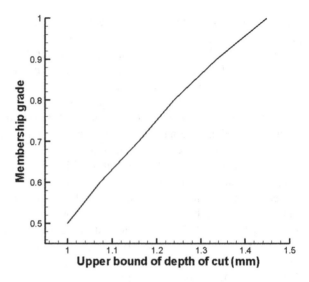

Fig. 6.8 Membership function for upper bound of depth of cut for two clamp design. With permission from Hazarika et al. [11]. Copyright [2010] Springer

bound of depth of cut at membership grades 0.5 and above for two clamp condition. High estimates of F_x, F_y, F_z and F_{clamp} at 0.5 membership grade are considered. Upper bound for depth of cut at 0.5 membership grade is 1 mm and at 1.0 membership grade, depth of cut can go up to 1.45 mm. Radius of the clamp r_{clamp} is found as 7 mm and minimum and maximum radius of curvature of the spherical clamp both are found to be 98 mm at 1 mm depth of cut. It is observed that the clamp parameters are same for both single clamp and double clamp design; only higher value of clamping force due to increased depth of cut is shared by two clamps.

The normal elastic deformation δ_n at the clamp-workpiece and locator-workpiece contact surface are calculated from Eq. (6.4) using worst case clamping force 643.50 N and highest locator reaction force 503.62 N. The values of δ_n are found to be 0.006 and 0.005 mm at the clamp-workpiece and locator-workpiece interface which are quite small.

Optimized locator and clamp layout is found considering the worst case clamping force 643.50 N at 0.5 mm depth of cut and 0.1 mm feed/tooth of the cutter. Layout optimization is formed as a constrained optimization problem and solved using nonlinear optimization technique FMINCON in MATLAB (Version 7). FMINCON uses sequential quadratic programming (SQP) to find a constrained minimum of a scalar function of several variables. It starts at an initial estimate and solves a quadratic sub-problem at each iteration. The solution of the sub-problem is used to find the search direction for an optimal solution. The design variables in the optimization problem are the locator and clamp positions. Table 6.1 shows the feasible region for positioning the locators and clamps on the workpiece surfaces. The optimized locator and clamp positions for minimized maximum norm of the locator reactions are given in Table 6.2. It is observed that the optimized locator and clamp layout gives a much lower value of the norm of the locator reactions (657.55 N) than the value (1,899.70 N) given by the initial locator and clamp layout.

Table 6.1 Feasible region for locators and clamps

Locator and clamp positions	Locator and clamp position constraints (mm)
$L_1 (x_1, y_1, z_1)$	$10 \leq x_1 \leq 60, 10 \leq y_1 \leq 50, z_1 = 0$
$L_2 (x_2, y_2, z_2)$	$10 \leq x_2 \leq 60, 10 \leq y_2 \leq 50, z_2 = 0$
$L_3 (x_3, y_3, z_3)$	$10 \leq x_3 \leq 60, 10 \leq y_3 \leq 50, z_3 = 0$
$L_4 (x_4, y_4, z_4)$	$10 \leq x_4 \leq 60, y_4 = 0, 10 \leq z_4 \leq 40$
$L_5 (x_5, y_5, z_5)$	$10 \leq x_5 \leq 60, y_5 = 0, 10 \leq z_5 \leq 40$
$L_6 (x_6, y_6, z_6)$	$x_6 = 0, 10 \leq y_6 \leq 50, 10 \leq z_6 \leq 40$
$C_1 (x_7, y_7, z_7)$	$x_7 = 70, 10 \leq y_7 \leq 50, 10 \leq z_7 \leq 40$
$C_2 (x_8, y_8, z_8)$	$10 \leq x_8 \leq 60, y_8 = 60, 10 \leq z_8 \leq 40$

With permission from Hazarika et al. [11]. Copyright [2010] Springer

Table 6.2 Optimized locator and clamp layout

Fixture elements	Initial locator and clamp layout	Optimized locator and clamp layout	Locator reactions R_i (N)
	(x, y, z) (mm)	(x, y, z) (mm)	
Locator L_1	(60, 30, 0)	(60, 18.12, 0)	0.73
Locator L_2	(45, 55, 0)	(60, 50, 0)	6.55
Locator L_3	(40, 35, 0)	(10, 10, 0)	37.72
Locator L_4	(10, 0, 10)	(10, 0, 26.65)	150.53
Locator L_5	(30, 0, 21)	(60, 0, 18.32)	393.22
Locator L_6	(0, 20, 22)	(0, 24.78, 19.60)	503.62
Clamp C_1	(70, 10, 20)	(70, 22.80, 25.82)	
Clamp C_2	(30, 60, 10)	(31.85, 60, 26.97)	

With permission from Hazarika et al. [11] and revised. Copyright [2010] Springer

6.6 Conclusion

Consideration of fixturing constraints in setup planning is inevitable for generation of a feasible and robust setup plan. In this chapter a methodology is presented for incorporating fixturing requirements into the setup planning expert system described in Chap. 4. The uncertainties associated with the work material, clamp material and clamping torque are considered by means of fuzzy arithmetic. The proposed setup planning system provides inputs to fixture designer in terms of recommended depth of cut and feed, fuzzy clamping forces, approximate optimal locator and clamp layout and sizes of the locators and clamps. Locators and clamps are designed based on machining and clamping forces. A strategy for finding the radius of curvature of the spherical locators and clamps is proposed for proper contact with the workpiece surface. The fixture designer can further optimize the fixture plan by taking these inputs from the setup planning module. Machining force, clamping force, recommended cutting parameters, initial fixture layout and proper size of the clamp/locator for applying the required clamping forces are some of the important issues considered in detail in this work. The methodology is explained with an example end milling process. The proposed methodology provides information to the fixture planner in order to enhance the feasibility of the fixture design. The information can be provided in offline as well as online mode. It is possible to integrate the proposed setup planning expert system with fixturing information generation module in a complete process planning system.

References

1. Attila R, Stampfer M, Imre S (2013) Fixture and setup planning and fixture configuration system. Procedia CIRP 7:228–233
2. Bansal S, Nagarajan S, Venkata RN (2008) An integrated fixture planning system for minimum tolerances. Int J Adv Manuf Technol 38:501–513
3. Boerma JR, Kals HJJ (1988) FIXES, a system for automatic selection of set-ups and design of fixtures. Ann CIRP 37:443–446
4. Boerma JR, Kals HJJ (1989) Fixture design with fixes: the automatic selection of positioning, clamping and supporting features for prismatic parts. Ann CIRP 38:399–402
5. Borgia S, Andrea Matta A, Tolio T (2013) STEP-NC compliant approach for setup planning problem on multiple fixture pallets. J Manuf Syst 32:781–791
6. Cai N, Wang L, Feng HY (2008) Adaptive setup planning of prismatic parts for machine tools with varying configurations. Int J Prod Res 46:571–594
7. Cecil J (2002) Computer aided fixture design: using information intensive function models in the development of automated fixture design system. SME J Manuf Syst 21:58–71
8. Chandrasekaran M, Muralidhar M, Murali Krishna CM, Dixit US (2010) Application of soft computing techniques in machining performance prediction and optimization: a literature review. Int J Adv Manuf Technol 46:445–464
9. Deng H, Melkote SN (2006) Determination of minimum clamping forces for dynamically stable fixturing. Int J Mach Tools Manuf 46:847–857
10. Gologlu C (2004) Machine capability and fixturing constraints-imposed automatic machining set-ups generation. J Mater Process Technol 148:83–92
11. Hazarika M, Dixit US, Deb S (2010) A setup planning methodology for prismatic parts considering fixturing aspects. Int J Adv Manuf Technol 51:1099–1199
12. Henriksen EK (1973) Jig and fixture design manual. Industrial Press, New York
13. Hu P, Tang K, Lee CH (2013) Global obstacle avoidance and minimum workpiece setups in five-axis machining. Comput Aided Des 45:1222–1237
14. Huang SH, Xu N (2003) Automatic set-up planning for metal cutting: an integrated methodology. Int J Prod Res 41:4339–4356
15. Johnson KL (1985) Contact mechanics. Cambridge University Press, Cambridge
16. Joneja A, Chang TC (1999) Setup and fixture planning in automated process planning systems. IIE Trans 31:653–665
17. Joshi PH (2000) Jigs and fixtures. Tata McGraw-Hill Publishing Company Limited, New Delhi
18. Kaya N, Ozturk F (2001) Algorithms for grouping machining operations and planning workpiece location under dynamic machining conditions. Int J Prod Res 39:3329–3351
19. Leonesio M, Tosatti LM, Pellegrinelli S, Valente A (2013) An integrated approach to support the joint design of machine tools and process planning. CIRP J Manufact Sci Technol 6:181–186
20. Li B, Melkote SN (2001) Fixture clamping force optimization and its impact on workpiece location accuracy. Int J Adv Manuf Technol 17:104–113

21. Li B, Melkote SN, Liang SY (2000) Analysis of reaction and minimum clamping force for machining fixtures with large contact areas. Int J Adv Manuf Technol 16:79–84
22. Liao YJG, Hu SJ (2000) Flexible multibody dynamics based fixture-workpiece analysis model for fixturing stability. Int J Mach Tools Manuf 40:343–362
23. Mehta NK (2004) Machine tool design and numerical control. Tata McGraw-Hill Publishing Company Limited, New Delhi
24. Sakurai H (1992) Automatic setup planning and fixture design in machining. J Manuf Syst 11:30–37
25. Shaw MC (2005) Metal cutting principles. Oxford University Press, New York
26. Stampfer M (2009) Automated setup and fixture planning system for box-shaped parts. Int J Adv Manuf Technol 45:540–552
27. Stephenson DA, Agapiou JS (2005) Metal cutting theory and practice. CRC Press, New York
28. Tseng YJ (1999) Fixturing design analysis for successive feature-based machining. Comput Ind 38:249–262
29. Young RI, Bell R (1991) Fixturing strategies and geometric queries in set-up planning. Int J Prod Res 29:537–550

Index

© The Author(s) 2015
M. Hazarika and U.S. Dixit, *Setup Planning for Machining*,
SpringerBriefs in Manufacturing and Surface Engineering,
DOI 10.1007/978-3-319-13320-1

Printed in the United States
By Bookmasters